Cambridge Tracts in Mathematics and Mathematical Physics

GENERAL EDITORS

P. HALL, F.R.S., AND F. SMITHIES, PH.D.

No. 51

INTEGRAL QUADRATIC FORMS

INTEGRAL
QUADRATIC FORMS

BY

G. L. WATSON, Ph.D.

*Lecturer, University College
London*

CAMBRIDGE
AT THE UNIVERSITY PRESS
1960

PUBLISHED BY
THE SYNDICS OF THE CAMBRIDGE UNIVERSITY PRESS

Bentley House, 200 Euston Road, London, N.W.1
American Branch, 32 East 57th Street, New York 22, N.Y.

CAMBRIDGE UNIVERSITY PRESS

1960

Printed in Great Britain at the University Press, Cambridge
(Brooke Crutchley, University Printer)

CONTENTS

PREFACE

The object of this tract is to give a modern, but fairly elementary, account of the theory of quadratic forms with integral coefficients and variables. The literature on this subject is voluminous but mostly makes difficult reading. One reason for this is the unsatisfactory treatment of the troublesome prime number 2 which results from Gauss's mistake of introducing binomial coefficients into the notation. Another is that the chronological order of development was far from logical; for example, Hasse in the 1920's was the first to recognize the importance of the simplest part of the theory, in which unrestricted rational transformations are allowed. And this depends on the theory of representation of zero, beginning with Meyer's theorem proved about 1872. Still another reason for the unsatisfactory state of the literature is that modern matrix notation, which is very desirable and was used about 100 years ago by Cayley, was not in general use till much later.

I assume that the reader is familiar with elementary number theory (divisibility, congruence, and primality). Some basic results of this theory, such as the possibility and uniqueness of factorization into primes, and the necessary and sufficient conditions for the solubility of a linear equation or congruence, are used without explicit mention, inferences depending on them being treated as obvious. It is possible (though I have tried to avoid it) that in the later chapters I may have treated in this way some of the results proved earlier (e.g. Theorems 1–4 and formulae (1)–(10), (14), (52), (74), and (76)–(85)). The only results from number theory that I use, but do not take for granted in this way, are Fermat's theorem and Dirichlet's theorem on primes in a progression. I prove what I need on quadratic residuacity (perhaps too sketchily), because it can be regarded as essentially the arithmetic of unary quadratic forms.

I assume scarcely any knowledge of other branches of mathematics, except for the rudiments of matrix algebra. In particular, I use practically no analysis, partly because analytical methods need a cumbrous notation for which I have no space, and partly because I find elementary methods very powerful.

The three problems on quadratic forms which I regard as fundamental are (i) that of equivalence, (ii) decomposition and (iii) the representation of integers. I do not include much that does not bear on these three problems. In particular, there are many properties of binary forms which are peculiar to them and do not extend to forms of higher rank. These I do not discuss; they are adequately treated elsewhere.

The notes on the chapters (p. 135) contain references to the literature, and are sometimes meant for a more sophisticated reader than the text. The bibliography makes no attempt at completeness, and is meant to be supplemented by the bibliographies of the books and papers listed in it, in particular by Dickson's *History*. I have not attempted to give a reference for every theorem I prove; thus the omission to mention a name implies no claim to originality. There are, however, some original results in the book, which I should have published separately had I not been engaged on it.

I hope that the book will be intelligible to beginners and serve as a desirable preparation for the study of recent deep work by Eichler, Kneser and others on quadratic forms over general rings.

If the proofs are found difficult I have two suggestions. First, it would sometimes make the proofs clearer to write out the matrix of a form in full; naturally I have been reluctant to do this. Secondly, the reader might omit, on a first reading (i) Theorems 34, 35, 67, 78, 80 and much of the proofs of Theorems 33, 36, 68, 74, 75, 81 (all this matter deals with special difficulties relating to the prime number 2); and (ii) sections 8–11 of Chapter 2 (which constitute a digression from the main plan).

I am grateful to friends, in particular Professor Oppenheim and Dr Chalk, who encouraged me to write it; to Professor Davenport and Dr Cassels, who read the proofs; and to Professor Hall, who read the typescript.

G. L. W.

LONDON
April 1959

NOTATION

1. We reserve the letters f, g, ϕ, ψ, σ for quadratic, and l for linear, forms (see Chapter 1, §1). Otherwise, small letters in ordinary type denote integers, except when the context clearly shows that a rational number, or very occasionally a real number, is meant. $\operatorname{sgn} a$ means $1, 0, -1$ according as $a >$, $=$, or < 0. $a \mid b$, $a \nmid b$ mean that a divides, does not divide, b. $a \equiv b \pmod{m}$ means that $m \mid a - b$. Primes are denoted by p, p_1, p', ...; and $p^u \| a$ ($p \| a$ if $u = 1$) means $p^u \mid a$ but $p^{u+1} \nmid a$. The phrase *greatest common divisor* is abbreviated to g.c.d.; and g.c.d.$(x_1, ..., x_n)$ means the g.c.d. of the integers $x_1, ..., x_n$.

2. Italic capitals denote square matrices, of any order n, unless the context shows that a rectangular matrix is meant. Bold-faced letters denote column vectors, or $n \times 1$ matrices. Matrix transposition is indicated by an accent ($'$). The identity matrix is denoted by I (or by I_n if there is any possibility of ambiguity regarding its rank n), and the $n \times n$ matrix with every element zero by a capital O (or O_n). The elements of a matrix are denoted by the corresponding small letters in ordinary type with suffixes, except that A (see Chapter 1, §2) has diagonal elements $2a_{ii}$. $m \mid M$ means $m \mid m_{ij}$ for $i, j = 1, ..., n$, and $M \equiv N \pmod{m}$ means $m \mid M - N$; similarly for vectors.

3. Brackets ( ) are used (i) in the usual ways and (ii) to enclose the elements of a row vector, e.g. $\mathbf{x}' = (x_1, ..., x_n)$, or the column vectors of a square matrix. Brackets { } are used to indicate that a vector, say $\mathbf{x} = \{x_1, ..., x_n\}$, written to save space as a row, is to be read as a column. Brackets [ ] are used (i) to denote the integral part of a real number (that is, for given real θ, $[\theta]$ is the unique integer with $[\theta] \leqslant \theta < [\theta] + 1$) and (ii) to enclose the elements of a diagonal matrix (or diagonal block matrix) (see Chapter 1, §2, (5), (6)). The signs | | are used (i) to denote the absolute value of a real number and (ii) to denote the determinant of a square matrix.

4. Many words, phrases and symbols used in special senses are listed on p. 143; each such word or phrase is set in small capitals where it is defined.

5. A list of works referred to is given on p. 141. They are quoted by giving the author's name and the year (with small letters a, b, ...to distinguish works in the same year). Selberg (1949) gives a proof (which though difficult is elementary in a somewhat technical sense) of the classical theorem of Dirichlet that if g.c.d. $(a, b) = 1$ and $b \neq 0$, then there exists a prime p congruent to a modulo b.

6. An asterisk before the statement of a theorem indicates that a proof is not given.

CHAPTER 1

INTRODUCTORY

1. Forms and their coefficients and variables

By a FORM is meant a homogeneous polynomial. Forms are classified as LINEAR, QUADRATIC, CUBIC, ..., according to their degree; we shall be concerned only incidentally with cases other than the quadratic, and so *form* means *quadratic form* if the degree is not mentioned. They are also classified as UNARY, BINARY, TERNARY, ..., n-ARY according as the number of variables is 1, 2, 3, ..., n. The variables x_1, x_2, ... may be regarded as the elements of the column vector, or $n \times 1$ matrix, $\mathbf{x} = \{x_1, ..., x_n\}$; also of its transpose, the row vector, or $1 \times n$ matrix,

$$\mathbf{x}' = (x_1, ..., x_n).$$

Thus, regarding its coefficients (say $c_1, c_2, ...$) also as elements of a vector, we may express a linear form concisely as a scalar product $\mathbf{c}'\mathbf{x}, = \mathbf{x}'\mathbf{c}$.

The coefficient of $x_i x_j$ in the quadratic form f will be denoted by $a_{ij}, = a_{ji}$. Thus we have

$$f = f(\mathbf{x}) = f(x_1, ..., x_n) = \sum_{1 \leqslant i \leqslant j \leqslant n} a_{ij} x_i x_j; \tag{1}$$

or alternatively

$$f = \tfrac{1}{2} x_1 (2a_{11} x_1 + a_{12} x_2 + ... + a_{1n} x_n) + ...$$
$$+ \tfrac{1}{2} x_n (a_{n1} x_1 + a_{n2} x_2 + ... + 2a_{nn} x_n). \tag{2}$$

The form f is said to be DISJOINT if it can be written as a sum $f_1 + f_2 + ...$, where f_1 involves only the x_i with $i \leqslant n_1$, f_2 only those with $n_1 < i \leqslant n_1 + n_2$, ..., for some set of positive integers n_1, n_2, ..., each less than n, whose sum is n. We shall adopt the convention that a sum $f_1 + f_2 + ...$ is assumed to be disjoint unless the contrary is indicated by mentioning the variables explicitly. In terms of the a_{ij}, f is disjoint if there exists n_1, with $0 < n_1 < n$, such that $i \leqslant n_1 < j$ implies $a_{ij} = 0$.

The a_{ii} are called the DIAGONAL COEFFICIENTS, and a_{11} the LEADING COEFFICIENT. The most highly disjoint forms are the DIAGONAL FORMS, that is, the forms whose only non-zero coefficients are the diagonal ones.

2. The matrix and discriminant of a quadratic form

It is clear that the matrix product $\mathbf{x}'M\mathbf{x}$ is, for every $n \times n$ matrix M (and $n \times 1$ $\mathbf{x}$), a quadratic form. (We may, when convenient, regard the product as a 1×1 matrix.) If m_{ij} is the general element of M, then for certain obvious values of $\mathbf{x}$ the form $\mathbf{x}'M\mathbf{x}$ assumes the values

$$m_{ii} \quad (i = 1, ..., n)$$

and $\qquad m_{ii} + m_{ij} + m_{ji} + m_{jj} \quad (i, j = 1, ..., n; \ i \neq j).$

It therefore vanishes identically in the x_i if and only if M is skew.

It follows that there is a unique symmetric matrix $A = A(f)$ such that the quadratic form f can be expressed as

$$f(\mathbf{x}) = \tfrac{1}{2}\mathbf{x}'A\mathbf{x}. \tag{3}$$

And by (2), this matrix A, which we call the (coefficient) MATRIX of f, has elements $2a_{ii}$ in the diagonal positions and a_{ij} elsewhere. From (3) we obtain the formula

$$f(\mathbf{x} + \mathbf{y}) = f(\mathbf{x}) + f(\mathbf{y}) + \mathbf{x}'A\mathbf{y}, \tag{4}$$

on noting that

$$\tfrac{1}{2}\mathbf{y}'A\mathbf{x} = \tfrac{1}{2}(\mathbf{y}'A\mathbf{x})' = \tfrac{1}{2}\mathbf{x}'A'\mathbf{y} = \tfrac{1}{2}\mathbf{x}'A\mathbf{y},$$

since a 1×1 matrix is symmetric, while A is symmetric by definition.

For a disjoint form $f_1 + f_2$ we have

$$A(f_1 + f_2) = [A(f_1), A(f_2)] = \begin{pmatrix} A(f_1) & \cdot \\ \cdot & A(f_2) \end{pmatrix}, \tag{5}$$

where each dot stands for an array of $n_1 n_2$ zeros; and similarly with three or more summands. In particular,

$$A = 2[a_{11}, ..., a_{nn}] \text{ if } f \text{ is diagonal}. \tag{6}$$

The DISCRIMINANT of f is

$$d = d(f) = \begin{cases} (-1)^{\frac{1}{2}n} |A| & (n \text{ even}), \\ \tfrac{1}{2}(-1)^{\frac{1}{2}n - \frac{1}{2}} |A| & (n \text{ odd}). \end{cases} \tag{7}$$

Notice that (7) gives

$$d = a_{11}, \quad a_{12}^2 - 4a_{11}a_{22} \quad \text{for } n = 1, 2, \tag{8}$$

$$d(f_1 + f_2) = \left\{ \begin{array}{ll} d(f_1)\,d(f_2) & (n_1 n_2 \text{ even}), \\ -4d(f_1)\,d(f_2) & (n_1 n_2 \text{ odd}), \end{array} \right\} \tag{9}$$

and

$$d = (-4)^{[\frac{1}{2}n]} a_{11} \ldots a_{nn} \quad (f \text{ diagonal}). \tag{10}$$

The definition is devised so as to give $d = 1$ when f is

$$x_1 x_2 + x_3 x_4 + \ldots,$$

with last term x_n^2 in case n is odd.

In the expansion of $|A|$, some terms contain diagonal factors $2a_{ii}$, while some others occur in pairs that are equal because of the symmetry of A. The only terms of neither of these kinds are products of $\frac{1}{2}n$ factors $-a_{ij}a_{ji}$, and so can occur only for even n. It follows that, in spite of the $\frac{1}{2}$ in $(7)_2$, d is a form, of degree n, in the $\frac{1}{2}n(n+1)$ variables a_{ij} with $i \leqslant j$, whose coefficients are integers. Thus when we restrict the a_{ij} to integral values, as we shall later, d will also be integral. We shall see later that without loss of generality we may restrict our attention to forms with $d \neq 0$.

3. The rank, signature and range of values of a quadratic form

From now on, let the variables x_i be real numbers, and let the coefficients a_{ij} be restricted so that the values of $f(\mathbf{x})$ are real; clearly this means that the a_{ij} must all be real. Consider the inequality $f(\mathbf{x}) \geqslant 0$. This inequality may hold for all real $\mathbf{x}$, and it certainly holds for $\mathbf{x}$ ($=\mathbf{0}$) satisfying the n linear equations $x_1 = \ldots = x_n = 0$. There exist therefore an integer $n^+ = n^+(f)$, with $0 \leqslant n^+ \leqslant n$, and n^+ linear forms l_j ($j = 1, \ldots, n^+$) in the x_i (each with real coefficients), such that the n^+ linear equations

$$l_1 = 0, \quad l_2 = 0, \quad \ldots \tag{11}$$

imply $f(\mathbf{x}) \geqslant 0$, while (if $n^+ > 0$) any set of simultaneous equations of this shape with fewer than n^+ members is consistent with $f(\mathbf{x}) < 0$. Similarly, we define $n^- = n^-(f)$ ($\geqslant 0$, $\leqslant n$) as the least possible number of equations, in a set of equations of the shape (11), which imply $f(\mathbf{x}) \leqslant 0$.

Now by the RANK of f is meant the number $n^+ + n^-$; while the SIGNATURE $s = s(f)$ is $n^- - n^+$. We shall see that the restriction $d \neq 0$ makes the rank of f equal to the number of variables, n. Assuming this, the numbers n^+, n^- are clearly $\frac{1}{2}(n \mp s)$, and we see that $|s| \leqslant n$, $s \equiv n \pmod 2$.

The form f is said to be INDEFINITE if it takes both positive and negative values, that is, if n^+ and n^- are both positive; which if the rank is n is the case if and only if $|s| < n$.

A form, of rank n, with $d \neq 0$, which is not indefinite is said to be DEFINITE. A definite form has n^+, $n^- = 0$, n or n, 0 and its values, for real $\mathbf{x} \neq \mathbf{0}$, are all positive in the first case, all negative in the second. In these cases respectively the form is POSITIVE (or positive-definite), NEGATIVE (or negative-definite). As an example of a form, with rank less than n, which is neither indefinite nor definite, take $n = 2$, $f = (x_1 + x_2)^2$, and note that $d = 0$.

Now, and henceforth unless the context clearly indicates that rational values are allowed, let $\mathbf{x}$ be integral, i.e. have integral elements x_i. We exclude the value $\mathbf{0}$ of $\mathbf{x}$ since trivially $f(\mathbf{0}) = 0$ for every f; and we say that $\mathbf{x}$ is PRIMITIVE if g.c.d. $(x_1, \ldots, x_n) = 1$. The form f is said to REPRESENT the number a if there is an integral $\mathbf{x} \neq \mathbf{0}$ such that $f(\mathbf{x}) = a$; and to REPRESENT a PROPERLY if there is a primitive $\mathbf{x}$ with $f(\mathbf{x}) = a$. f is said to be a ZERO FORM if it represents 0. Notice that f is a zero form if there is a rational $\mathbf{t} \neq \mathbf{0}$ with $f(\mathbf{t}) = 0$, and that a zero form represents zero properly. To prove these two assertions, note that if $\mathbf{t}$ is rational and $f(\mathbf{t}) = 0$ then $\mathbf{x} = \theta\mathbf{t}$ is integral and primitive for some rational θ unless $\mathbf{t} = \mathbf{0}$, and $f(\mathbf{x}) = \theta^2 f(\mathbf{t}) = 0$. Notice also that f represents properly all its diagonal coefficients, and all the integers

$$a_{ii} + a_{jj} + a_{ij}, \quad 1 \leqslant i < j \leqslant n. \tag{12}$$

(Take $\mathbf{x}$ to be (i) a base vector, (ii) the sum of two base vectors.)

It follows from this last remark that f is INTEGRAL, that is, represents integers only, if and only if the a_{ij} are all integral; and this we shall henceforth assume to be the case, except occasionally when otherwise stated. An integral form is said to be PRIMITIVE if the g.c.d. of all its values is 1; this is clearly so if and only if the g.c.d. of the a_{ij} is 1. In any case, this g.c.d. is called the DIVISOR of f. If f is primitive and m an integer $\neq 0$, then there exists an integral $\mathbf{x}$ (whence obviously a primitive $\mathbf{x}$) such that $f(\mathbf{x})$ is prime to m. When m has just one prime factor (p) this statement is obvious (the a_{ii} and the integers (12) are not all divisible by p). To prove it generally, let $p_1, p_2, \ldots$ be the distinct primes dividing m, and find integral $\mathbf{x}_1, \mathbf{x}_2, \ldots$ with $p_1 \nmid f(\mathbf{x}_1)$, $p_2 \nmid f(\mathbf{x}_2)$, $\ldots$. Supposing without loss of generality that $m = p_1 p_2, \ldots$ (i.e. that m is square-free), we see that $\mathbf{x} = m p_1^{-1} \mathbf{x}_1 + m p_2^{-1} \mathbf{x}_2 + \ldots$ does what is wanted.

4. Change of variables

We shall make frequent use of LINEAR TRANSFORMATIONS

$$\mathbf{x} = R\mathbf{y}, \quad R \text{ rational}, \quad |R| \neq 0. \tag{13}$$

The form $f(R\mathbf{y})$, in the new variables y_i, arising from f by the transformation (13) is denoted by f^R. When there is no risk of confusion we shall often keep to the original notation for the variables, that is, write f^R as $f^R(\mathbf{x}) = f(R\mathbf{x})$, and regard it as obtained from f by the SUBSTITUTION $\mathbf{x} \to R\mathbf{x}$. The form f' is said to be RATIONALLY RELATED to f if there exists an R satisfying $(13)_2$, $(13)_3$ such that $f'(\mathbf{x}) = f^R(\mathbf{x})$, identically in $\mathbf{x}$.

It is clear that the rank and signature of a form f are unaltered by any transformation (13), that is, they have the same values for f^R as for f. The property of representing or not representing zero is also unaffected (since $\mathbf{x}$ is rational if and only if $\mathbf{y}$ is so; see the remark following the definition of a zero form). The effect on the matrix and discriminant is given by

$$A(f^R) = R'A(f)R, \quad d(f^R) = |R|^2 d(f), \tag{14}$$

since $f^R(\mathbf{x}) = \frac{1}{2}\mathbf{x}'R'AR\mathbf{x}$ and $R'AR$ is symmetric.

An important special case of (13) is the EQUIVALENCE TRANSFORMATION

$$\mathbf{x} = T\mathbf{y}, \quad T \text{ integral}, \quad |T| = \pm 1. \tag{15}$$

With T satisfying these conditions, T^{-1} exists; it has the value $|T|^{-1} \operatorname{adj} T$, and it satisfies the same conditions as T. It follows that $\mathbf{x} = T\mathbf{y}$ sets up a one-to-one correspondence between integral $\mathbf{x}$ and integral $\mathbf{y}$, hence also between primitive $\mathbf{x}$ and primitive $\mathbf{y}$. Because of this property, two forms which are EQUIVALENT to each other, that is, are related by an equivalence transformation, have all essential properties in common. In particular, they represent the same integers, represent the same integers properly, and have the same rank, signature and discriminant (see (14)).

It is often necessary to construct an equivalence transformation whose matrix T has some convenient property. As a first step in this direction we prove:

THEOREM 1. *Let* $\mathbf{t} = \{t_1, ..., t_n\}$ *be primitive. Then there exists an* $n \times n$ *matrix* T *with integral elements whose determinant is* ± 1 *and whose first column vector is* $\mathbf{t}$.

Every form f which represents properly an integer a is equivalent to one with a as its leading coefficient.

Proof. The first assertion is trivial for $n = 1$; and the condition that $\mathbf{t}$ be primitive is clearly necessary. For $n = 2$, we can choose integers z_1, z_2 so that $t_1 z_2 - t_2 z_1 = 1$; then we take $T = (\mathbf{t}, \mathbf{z})$, $\mathbf{z} = \{z_1, z_2\}$. For $n \geqslant 3$, we use induction on n. We may write $\mathbf{t} = \{t_1, h\mathbf{z}\}$, with $\{t_1, h\}$ and $\mathbf{z}$ primitive. By the inductive hypothesis there is an integral 2×2 matrix, say V, with $\{t_1, h\}$ as its first column, and an integral $(n-1) \times (n-1)$ matrix, say Z, with first column $\mathbf{z}$, each of $|V|$, $|Z|$, being ± 1. Now we put

$$T = [1, Z][V, I_{n-2}], \quad |T| = |Z||V| = \pm 1,$$

and we evaluate the first column of T; it is $T\{1, \mathbf{0}\}$, which is

$$[1, Z][V, I_{n-2}]\{1, 0, \mathbf{0}\} = [1, Z]\{t_1, h, \mathbf{0}\} = \{t_1, h\mathbf{z}\} = \mathbf{t}.$$

Thus the first assertion is proved; it is clear that except in the trivial case $n = 1$ we can have $|T| = 1$.

For the second assertion, apply the first, with $\mathbf{t}$ such that $f(\mathbf{t}) = a$. The leading coefficient of the equivalent form

$$f^T = \tfrac{1}{2}\mathbf{x}'T'AT\mathbf{x}$$

is $\quad\quad \tfrac{1}{2}(1, \mathbf{0}')\,T'AT\{1, \mathbf{0}\} = \tfrac{1}{2}\mathbf{t}'A\mathbf{t} = f(\mathbf{t}) = a.$

A corollary is that an integral form with $d = 0$ is equivalent to an $(n-1)$-ary form, whence there is no loss of generality in assuming $d \neq 0$. To prove this, note that $d = 0$ gives $|A| = 0$. Hence there is a vector $\mathbf{t}$ with $\mathbf{t} \neq \mathbf{0}$, $A\mathbf{t} = \mathbf{0}$. The well-known construction for such a $\mathbf{t}$ shows that it can be taken (for rational

A) to be rational; whence multiplying it by a suitable scalar we may suppose it integral and primitive. Now apply the first part of Theorem 1. The first column of the matrix $T'AT$ of the form f^T equivalent to f is $T'AT\{1, 0\} = T'A\mathbf{t} = \mathbf{0}$. This shows that $f^T(\mathbf{y})$ is independent of y_1; we therefore regard it as an $(n-1)$-ary form in $y_2, \dots, y_n$.

5. Relations between forms

We shall use many different relations between forms. Each such relation, say $\mathcal{R}$, will be defined by saying that $f\mathcal{R}f'$ means that it is possible to pass from the form f to f' by suitably chosen operations satisfying certain conditions. These operations will usually be linear transformations. Sometimes, however, the operation of multiplying the form by a rational number will be allowed; this operation, for example, enables us to pass from an imprimitive form to a primitive one, of which it is a multiple. And in investigating congruence properties, to a modulus m, we may allow the operation of adding arbitrary multiples of m to the coefficients.

We shall be concerned with relations $\mathcal{R}$ with the REFLEXIVE, SYMMETRIC and TRANSITIVE properties:

$$f\mathcal{R}f \text{ (all } f); \quad f\mathcal{R}f' \text{ implies } f'\mathcal{R}f;$$

and $\qquad\qquad f\mathcal{R}f'\mathcal{R}f'' \text{ implies } f\mathcal{R}f''. \qquad\qquad (16)$

The relation of equivalence defined in the last section will be denoted as usual by $\sim$; that is, $f \sim f'$ means $f^T = f'$, with T satisfying $(15)_2$ and $(15)_3$. Now these conditions hold for $T = I$, and if they hold for $T = T_1$ and for $T = T_2$ then they also hold for $T = T_1^{-1}$ and for $T = T_1 T_2$. It follows that equivalence has the properties (16); and so, by a similar argument, has rational relatedness.

Any relation with the properties (16) determines a classification of all forms into 'families', such that $f\mathcal{R}f'$ if and only if f, f' are in the same family. In the case of equivalence, the family is called a CLASS. The family breaks up into smaller families when we pass to another relation, say $\mathcal{S}$, which is narrower than $\mathcal{R}$ (that is, is such that $f\mathcal{S}f'$ implies $f\mathcal{R}f'$, but not conversely).

If T in (15) is suitably chosen the y_i can be any permutation of the x_i; in other words, the substitution $\mathbf{x} \to T\mathbf{x}$ can be chosen so as to permute the x_i in any way. We shall use only relations $\mathscr{R}$ so defined that every such permutation is among the operations allowed. Hence, in particular, we shall always have

$$f_1 + f_2 \mathscr{R} f_2 + f_1. \tag{17}$$

Usually, but not always, $\mathscr{R}$ will have also the additive property

$$f_2 \mathscr{R} f'_2 \text{ implies } f_1 + f_2 \mathscr{R} f_1 + f'_2. \tag{18}$$

If any property of quadratic forms is such that whenever f has the property then every f' with $f\mathscr{R}f'$ has it too, then the property is said to be INVARIANT under $\mathscr{R}$. For example, the property of representing zero is, as we have noticed, rationally invariant, that is, invariant under rational relatedness. If a function, say $\mu(f)$, of the coefficients of f is such that $f\mathscr{R}f'$ implies $\mu(f) = \mu(f')$, then $\mu(f)$ is said to be (an) invariant under $\mathscr{R}$. (These two concepts are essentially the same, as may be seen by defining $\mu(f)$ to be 1, -1 according as f has or has not a certain property.) We investigate such a property by seeking a form f' which, on the one hand, satisfies $f\mathscr{R}f'$, for some given f, and on the other hand is in some sense as simple as possible. The transitive property (16) enables us to do this by successive steps; and the desired simplicity usually consists either in having small coefficients or in disjointness.

In the latter case, we seek a DECOMPOSITION under $\mathscr{R}$ of a given form f; that is, we try to find $f_1, f_2, \ldots$ so that

$$f\mathscr{R}f_1 + f_2 + \ldots. \tag{19}$$

The properties (17) and (18) permit us to permute the summands in such a decomposition; or to REFINE it, if possible, by further decomposing any summand. We may also COARSEN it by bracketing together two or more summands. Obviously the decomposition has to be coarser when the relation is narrower, and so coarsening may be a convenient preliminary to modifying the decomposition so as to make it hold under a narrower relation.

6. Rational diagonalization

We give an easy illustration of the notion of decomposition. It is convenient to say that f_1 SPLITS OFF from f, under $\mathscr{R}$, if an integral f_2 can be found so that $f\mathscr{R}f_1 + f_2$.

THEOREM 2. *Let* $f_1 = f(x_1, ..., x_{n_1}, 0, ..., 0)$, *for some* n_1 *with* $0 < n_1 < n$; *and suppose* $d(f_1) \neq 0$. *Then* f_1 *splits off rationally* (*i.e. under rational relatedness) from* f.

Proof. It is easy to deduce the general result from the special case $n_1 = 1$. We use the transformation

$$x_1 = y_1 - a_{12}y_1 - ... - a_{1n}y_n, \quad x_i = 2a_{11}y_i \quad (i = 2, ..., n). \quad (20)$$

Plainly this transformation takes f into a form with leading coefficient a_{11} and with no terms in $y_1y_2, ..., y_1y_n$. As its coefficients are integral, it must give an integral f_2. The determinant of the transformation is $(2a_{11})^{n-1} \neq 0$, since $a_{11} = d(f_1)$ is assumed not to vanish.

The general case can, however, be dealt with directly. By straightforward matrix multiplication we see that $f^R = f_1 + f_2$, with integral f_2, for R defined by

$$R = \begin{pmatrix} I & -(\operatorname{adj} A_1)B \\ . & |A_1|I \end{pmatrix}, \quad A = \begin{pmatrix} A_1 & B \\ B' & A_2 \end{pmatrix}, \quad (21)$$

with an obvious partitioning. Here $|R| = |A_1|^{n-n_1}$, where

$$|A_1| = |A(f_1)| = \pm d(f_1) \text{ or } \pm 2d(f_1) \neq 0.$$

We deduce:

THEOREM 3. *Every form* f *is rationally related to a diagonal form, which may be chosen to have leading coefficient* $a \neq 0$, *if* f *represents* aq^2 *for some* $q \neq 0$.

Proof. By Theorem 2, we can split off a unary form, $a_{11}x_1^2$, if $a_{11} \neq 0$. By putting θx_1 for x_1, we can split off $a_{11}\theta^2 x_1^2$ for any rational θ for which $a_{11}\theta^2$ is integral. Appealing to Theorem 1, we see easily that we can split off ax_1^2 under the stated conditions on a. We complete the diagonalization inductively, using the notion of refinement explained in the last section. It should be noted that if f represents no non-zero integer it is trivially diagonal, and, moreover, has discriminant 0.

This result leads to a formula for the signature, s, which in the literature is generally used to define s. It also gives us a result previously mentioned regarding the rank.

THEOREM 4. *The rank of a form with discriminant $\neq 0$ is equal to the number of variables; and the signature of a diagonal form f is*
$$\sum_{1 \leqslant i \leqslant n} \operatorname{sgn} a_{ii}.$$

Proof. The rank being obviously a rational invariant, we may suppose the form f to be diagonal for both parts of the theorem. We shall assume $d \neq 0$ in both parts, though the second assertion is true without this restriction; and the rank of f is equal to that of its matrix A in all cases, being $\Sigma \operatorname{sgn} |a_{ii}|$ in the diagonal case. Note also that the property $d \neq 0$ is rationally invariant, by $(13)_3$ and $(14)_2$. Plainly a diagonal form has $d \neq 0$ if and only if no a_{ii} is 0.

Now suppose that there are n_1 positive and n_2 negative a_{ii}, with $n_1 \geqslant 0$, $n_2 \geqslant 0$, $n_1 + n_2 = n$. Both assertions follow if we show that the integers n^+, n^- defined in §3 are equal to n_2, n_1 respectively. Plainly $n^+ \leqslant n_2$, $n^- \leqslant n_1$. For $f(\mathbf{x}) \geqslant 0$ is implied by the n_2 equations $x_i = 0$ for the i with $a_{ii} < 0$; and $f(\mathbf{x}) \leqslant 0$ is implied by the n_1 equations

$$x_i = 0 \quad \text{for} \quad a_{ii} > 0. \tag{22}$$

Now (22) implies $f(\mathbf{x}) = \sum_{a_{ii} < 0} a_{ii} x_i^2 \leqslant 0$, with equality only if $\mathbf{x} = \mathbf{0}$. Since the equations (11) imply $f(\mathbf{x}) \geqslant 0$, (11) and (22) together imply $\mathbf{x} = \mathbf{0}$. This is possible only if they together contain at least n equations. Hence $n^+ + n_1 \geqslant n$, $n^+ \geqslant n_2$. So we must have $n^+ = n_2$, and $n^- = n_1$ is proved in the same way.

Since, as previously remarked (§3), n^+, $n^- = \frac{1}{2}(n \mp s)$, we have $n_1, n_2 = \frac{1}{2}(n \pm s)$, and so $n_1 - n_2 = \Sigma \operatorname{sgn} a_{ii} = s$. This proves the theorem and gives the following formulae, which show that n and $\operatorname{sgn} d$ determine s modulo 4:

$$|s| \leqslant n, \quad s \equiv n \pmod{2}, \quad (-1)^{[\frac{1}{2}s]} = \operatorname{sgn} d. \tag{23}$$

To obtain $(23)_3$ we use the rational invariance of s and $\operatorname{sgn} d$, and (10), and note that by (10)

$$\operatorname{sgn} d = (-1)^{[\frac{1}{2}n] - \frac{1}{2}(n-s)} = (-1)^{[\frac{1}{2}s]}.$$

7. Rational automorphs

By an AUTOMORPH of an n-ary form f (or of its matrix A) is meant an $n \times n$ matrix S such that

$$S'AS = A, \quad \text{implying } |S| = \pm 1, \tag{24}$$

since $|S|^2 |A| = |A| = \pm d$ or $\pm 2d \neq 0$. If S satisfies (24) then the transformation $\mathbf{x} = S\mathbf{y}$, or the substitution $\mathbf{x} \to S\mathbf{x}$, leaves f unaltered, since $f^S(\mathbf{x}) = \frac{1}{2}\mathbf{x}'S'AS\mathbf{x} = f(\mathbf{x})$ identically. The converse is true since this identity implies that the symmetric matrix $S'AS - A$ is also skew.

For any $\mathbf{t}$ with $f(\mathbf{t}) \neq 0$, define

$$U(\mathbf{t}) = U(\mathbf{t}, f) = U(\mathbf{t}, A) = I - \mathbf{tt}'A/f(\mathbf{t}). \tag{25}$$

Then $U(\mathbf{t})$ is an automorph of f; for

$$f^2(\mathbf{t})\, U'(\mathbf{t}) A U(\mathbf{t}) = f^2(\mathbf{t}) A - 2f(\mathbf{t}) A\mathbf{tt}'A + A\mathbf{t}(\mathbf{t}'A\mathbf{t})\mathbf{t}'A = f^2(\mathbf{t}) A.$$

We shall call such an automorph a REFLEXION. It is clear that $U(\mathbf{t})$ is rational if and only if the ratios of the t_i are rational; and since obviously $U(\theta\mathbf{t}) = U(\mathbf{t})$ for $\theta \neq 0$, we lose nothing by considering, if convenient, only integral, or primitive, $\mathbf{t}$. It is clear from (25) that

$$U(\mathbf{t})\mathbf{t} = -\mathbf{t}, \quad U(\mathbf{t})\mathbf{x} = \mathbf{x} \quad \text{if} \quad \mathbf{t}'A\mathbf{x} = 0. \tag{26}$$

It may be worth while to remark that the obvious automorph

$$x_1 \to -a_{11}^{-1}(a_{11}x_1 + a_{12}x_2 + \ldots + a_{1n}x_n), \quad x_i \to x_i \quad (i \neq 1),$$

is just the case $\mathbf{t} = \{1, \mathbf{0}\}$ of $\mathbf{x} \to U(\mathbf{t})\mathbf{x}$.

THEOREM 5. *If* $f(\mathbf{y}) = f(\mathbf{z}) \neq 0$ *there exists a rational automorph* S *of* f *with* $S\mathbf{y} = \mathbf{z}$, $S\mathbf{z} = \mathbf{y}$.

Proof. We take $S = U(\mathbf{y} - \mathbf{z})$ if $U(\mathbf{y} - \mathbf{z})$ is defined, that is, if $f(\mathbf{y} - \mathbf{z}) \neq 0$. For with this choice of S we have

$$S(\mathbf{y} - \mathbf{z}) = \mathbf{z} - \mathbf{y}, \quad S(\mathbf{y} + \mathbf{z}) = \mathbf{y} + \mathbf{z},$$

by (26) and

$$(\mathbf{y}' - \mathbf{z}') A(\mathbf{y} + \mathbf{z}) = 2f(\mathbf{y}) - 2f(\mathbf{z}) + \mathbf{y}'A\mathbf{z} - \mathbf{z}'A\mathbf{y} = 0.$$

By a similar argument, we may take

$$S = -U(\mathbf{y} + \mathbf{z}) \quad \text{if} \quad f(\mathbf{y} + \mathbf{z}) \neq 0.$$

If both choices fail then, using (4), we find

$$f(\mathbf{y} + \mathbf{z}) + f(\mathbf{y} - \mathbf{z}) = 2f(\mathbf{y}) + 2f(\mathbf{z}) = 0,$$

contrary to hypothesis.

An important consequence of Theorem 5 is (cf. (18))

THEOREM 6. *If $f_1 + f_2$ is rationally related to $f_1 + f_2'$, then f_2 is rationally related to f_2'.*

Proof. By Theorem 3 we may suppose f_1 diagonal. There is thus an easy induction on its rank n_1, once we have dealt with the case $n_1 = 1$. So we may suppose, for an integer $a \neq 0$ and rational R with $|R| \neq 0$, that

$$f = ax_1^2 + f_2, \quad f^R = ax_1^2 + f_2'.$$

Hence also, for every rational automorph S of f, we have

$$f^{SR} = (ax_1^2 + f_2)^{SR} = ax_1^2 + f_2'. \tag{27}$$

Now if SR is of the shape $[1, Q]$, (27) gives us the desired result $f_2^Q = f_2'$.

It suffices therefore to find an S such that SR has first column $\{1, \mathbf{0}\}$ and first row $(1, \mathbf{0}')$, that is, with

$$SR\{1, \mathbf{0}\} = \{1, \mathbf{0}\}, \quad (1, \mathbf{0}')SR = (1, \mathbf{0}'). \tag{28}$$

Now if $(28)_1$ holds but $(28)_2$ fails, then the terms in $x_1 x_2, \ldots, x_1 x_n$ in $(ax_1^2 + f_2)^{SR}$ cannot cancel out, and so (27) is contradicted. We have therefore only to solve $(28)_1$. This may be written $S\mathbf{y} = \mathbf{z}$, where $\mathbf{z} = \{1, \mathbf{0}\}$ and $\mathbf{y} = R\{1, \mathbf{0}\}$. ($\mathbf{y}$ is in general not integral, but this is immaterial.)

The existence of a suitable S now follows from Theorem 5 on noting that $f(\mathbf{z}) = a$ and $f(\mathbf{y}) = f^R(\mathbf{z}) = a$, and the proof is complete.

Theorem 6 has no analogue for most of the other relations we shall use. In case of equivalence, a simple counter-example is

$$x_1^2 + x_2^2 - x_3^2 = 2(x_1 - x_3)(x_2 + x_3) + (x_1 - x_2 - x_3)^2$$
$$\sim x_1^2 + 2x_2 x_3, \tag{29}$$

whereas obviously $x_2^2 - x_3^2 \nsim 2x_2 x_3$.

Chapter 2

REDUCTION

1. The problem of reduction; the minimum

The problem of reduction is to transform a given form, if possible, into a simpler one equivalent to it; here *simpler* means having smaller coefficients, in a sense which we must make precise. Alternatively, we may regard it as the problem of picking out, from a given class (that of all forms equivalent to some given one), the forms with smallest coefficients. We shall do this, in various ways, by saying that a form f is reduced if its coefficients satisfy certain conditions. These conditions (which may depend on n, s) are so devised as to imply that the reduced forms in any class are the simplest ones in that class. It is essential to frame the reduction conditions so that every class contains at least one reduced form. It is desirable that each class should contain exactly one reduced form. It is desirable also that to each of the reduction conditions there should correspond an equivalence transformation taking every form not satisfying that condition into a simpler one. If so, then we have a method by which any given form can be taken into a reduced one in a finite number of steps. (The coefficients, being integral, cannot infinitely often be made smaller.)

We shall begin, at first, by making the leading coefficient as small as possible but not 0. This leads us to define the MINIMUM of f, in symbols $\min f$, as the least positive value of $|f(\mathbf{x})|$ for integral $\mathbf{x}$. (Since every set of positive integers has a least member, this definition breaks down only in the trivial case in which f is identically 0, giving $d = 0$.) Clearly f represents properly at least one of $\pm \min f$; for if $|f(h\mathbf{y})| = \min f$, $h > 1$, $\mathbf{y}$ integral, then f represents one of $\pm h^{-2} \min f$ and so $|f|$ takes a value which is positive but less than $\min f$, contradicting the definition. So by Theorem 1 we can transform every f into an equivalent form with leading coefficient $\pm \min f$. It is generally convenient to assume that the given form f, which we seek to reduce, already satisfies this condition, that is, that $\min f = |a_{11}|$. This assumption simplifies the notation, and is permissible because of the transitive property of equivalence and the obvious invariance under equivalence of the minimum.

Having made a_{11} small, we proceed further by transformations of two types, each leaving a_{11} unaltered. First we have transformations which do not affect x_1, that is, which have matrices of the shape $[1, T]$, T integral, $(n-1) \times (n-1)$, and with determinant ± 1. Next we have the PARALLEL TRANSFORMATIONS

$$x_1 = y_1 + h_2 y_2 + \ldots + h_n y_n, \quad x_i = y_i \text{ for } i > 1. \qquad (30)$$

By suitable choice of the integers h_j, (30) takes f into an equivalent form whose coefficients a'_{ij} satisfy

$$a'_{11} = a_{11}, \quad a'_{1j} = a_{1j} + 2h_j a_{11}, \quad j = 2, \ldots, n$$

so that the a'_{1j} with $j > 1$ can be taken to lie in any desired intervals, each of length $2|a_{11}|$, provided $a_{11} \neq 0$. Again it simplifies the notation to assume that f itself has coefficients $a_{12}, \ldots, a_{1n}$ lying in such intervals.

2. Binary forms

We consider in this section binary forms not representing zero. First we deal with positive forms. By a REDUCED POSITIVE BINARY FORM is meant one with

$$a_{11} > 0, \quad 0 \leqslant a_{12} \leqslant a_{11} \leqslant a_{22}. \qquad (31)$$

THEOREM 7. *Every positive binary form is equivalent to a unique reduced form. A reduced positive binary form satisfies*

$$\min f = a_{11}, \quad |d| = -d \geqslant 3 a_{11} a_{22} \geqslant 3 \min^2 f. \qquad (32)$$

Proof. By a suitable parallel transformation we may suppose $-a_{11} \leqslant a_{12} \leqslant a_{11}$, a_{11} being necessarily positive. Putting if necessary $-x_2$ for x_2, we may suppose $a_{12} \geqslant 0$, and then we have $0 \leqslant a_{12} \leqslant a_{11}$. Now if we assume $(32)_1$ we have $a_{22} \geqslant \min f = a_{11}$, and so f is already reduced. Alternatively, if $a_{22} < a_{11}$ we take the form into one with smaller leading coefficient by interchanging the variables; we cannot do this infinitely often, and so

we ultimately obtain a reduced form. $(32)_2$ follows from (31), since
$$-d = 4a_{11}a_{22}-a_{12}^2 \geqslant 4a_{11}a_{22}-a_{11}^2;$$

and strict inequality holds in each place in (32) except for the form $a_{11}(x_1^2+x_1x_2+x_2^2)$.

It remains to prove the uniqueness. From (31) we have

$$f(\mathbf{x}) \geqslant a_{11}(x_1+\tfrac{1}{2}a_{11}^{-1}a_{12}x_2)^2+\tfrac{3}{4}a_{22}x_2^2 \geqslant 3a_{22} > a_{11}+a_{22}$$

if $|x_2| \geqslant 2$. If $x_2 = \pm 1$ we have $f > a_{11}+a_{22}$ except for $x_1 = 0$, ± 1. It follows that the three smallest values, for primitive $\mathbf{x}$, of f are a_{11}, a_{22}, $a_{11}-a_{12}+a_{22}$. Clearly therefore two equivalent reduced forms must have the same a_{11}, a_{22}, a_{12}, and so must be identically equal. This completes the proof. We see that we need not begin by assuming $(32)_1$, as it follows from (31).

There is no need to consider negative forms, since $-f$ is positive if f is negative.

THEOREM 8. *Every indefinite binary form not representing zero is equivalent to one satisfying*

$$0 \leqslant a_{12} \leqslant |a_{11}| \leqslant |a_{22}|, \quad a_{11}a_{22} < 0; \tag{33}$$

and also $$d \geqslant 8a_{11}^2 \tag{34}$$

unless the form is equivalent to a multiple of

$$x_1^2+x_1x_2-x_2^2. \tag{35}$$

Proof. The argument for $(33)_1$ is the same as for $(31)_2$; we may, but need not, begin by assuming $\min f = |a_{11}|$. If $a_{11}a_{22} \geqslant 0$ it is easy to see that the form is definite or represents zero, contrary to hypothesis.

To obtain (34), we consider the numbers

$$f(1, \pm 1) = a_{11} \pm a_{12}+a_{22}.$$

As each of these is the leading coefficient of a form equivalent to f, we may suppose that the modulus of the smaller, which is $\pm(|a_{11}|+|a_{12}|-|a_{22}|)$, by (33), is not less than $|a_{11}|$. Hence either $|a_{22}| \leqslant |a_{12}|$, which with (33) gives $a_{11} = \pm a_{12} = -a_{22}$, or $|a_{22}| \geqslant 2|a_{11}|+|a_{12}|$. In the latter case we have

$$d = 4|a_{11}a_{22}|+a_{12}^2 \geqslant 8a_{11}^2.$$

A form satisfying (33) may for the moment be called reduced, though this is not the only, nor an entirely satisfactory, definition.

3. Representation of integers by binary forms

A binary zero form, which we may take to be $x_2(a_{12}x_1 + a_{22}x_2)$, obviously represents an integer a if and only if $a = a_1 a_2$, with $a_2 \equiv a_{22}a_1 \pmod{a_{12}}$. In particular, if $d = a_{12}^2 = 1$, the form is equivalent to $x_1 x_2$, and represents all integers. Hence for the rest of this section we consider only binary forms not representing zero. It is clearly sufficient to consider the representation of positive integers a by forms f which are either indefinite or positive; so we exclude negative forms. It is simpler to consider only proper representation; results on improper representation follow immediately.

If the integer $a > 0$ is represented properly by any integral binary form with given discriminant d then it is obviously (by Theorem 1) represented by a form of the shape

$$ax_1^2 + bx_1x_2 + \tfrac{1}{4}a^{-1}(b^2 - d)\,x_2^2. \tag{36}$$

We may suppose $0 \leqslant b \leqslant a$, and so the necessary and sufficient condition for such a representation is the existence of an integer b satisfying

$$b^2 \equiv d \pmod{4a}, \quad 0 \leqslant b \leqslant a. \tag{37}$$

We shall see later that this is a severe restriction, which most a do not satisfy; but for a which do satisfy it, there may be several forms (36), corresponding to different solutions of (37).

Using the results of the last section to reduce all possible forms (36), with given a, d, we find exactly which reduced forms with discriminant d represent a. The case in which there is only one (indefinite or positive) reduced form with discriminant d is of special interest. If so, that form clearly represents properly every $a > 0$ for which (37) is soluble. In particular, if $d = -4$ we see that (31) and (32) imply $a_{11} = a_{22} = 1$, $a_{12} = 0$, and the only reduced form is $x_1^2 + x_2^2$. This form therefore represents properly every $a > 0$ for which $b^2 \equiv -4 \pmod{4a}$ is soluble. Using the results of the next chapter, we obtain the classical result that a positive integer is the sum of two coprime squares if and only if it is divisible neither by 4 nor by any prime congruent to -1 modulo 4. There are similar results for $d = -3, -7, -8, -11$.

If there are two or more reduced forms we may still be able to obtain elementary results of this type. For example, there are two reduced positive forms with $d = -12$, namely, $x_1^2 + 3x_2^2$ and $2(x_1^2 + x_1 x_2 + x_2^2) = f_1, f_2$, say. Proper representation of a by f_1, f_2 clearly implies $a \not\equiv 2$, $a' \equiv 2 \pmod{4}$. Hence $x_1^2 + 3x_2^2$ represents properly every positive $a \not\equiv 2 \pmod{4}$ and such that $b^2 \equiv -12 \pmod{4a}$ (or $b'^2 \equiv -3 \pmod{a}$) is soluble (and no other integer). Again, suppose $d = 8$; then there are two reduced forms $\pm(x_1^2 - 2x_2^2)$, but they are equivalent to each other since they represent ∓ 1 respectively. So each represents properly all a for which $b^2 \equiv 8 \pmod{4a}$ is soluble. For another example, take $d = -20$. There are two reduced forms,

$$x_1^2 + 5x_2^2 \quad \text{and} \quad 2x_1^2 + 2x_1 x_2 + 3x_2^2 \sim 2x_1^2 + 10x_1 x_2 + 15x_2^2.$$

The first can represent a properly only if $a \equiv \pm 1 \pmod 5$ or $\pm 5 \pmod{25}$, the second only if $a \equiv \pm 2 \pmod 5$ or $\pm 10 \pmod{25}$. Hence we know exactly which integers each represents.

When all these arguments fail, we shall pursue the problem no further, as it requires methods which would take us too far afield. For the simplest example, consider the two reduced positive forms, $x_1^2 + x_1 x_2 + 6x_2^2$, $2x_1^2 + x_1 x_2 + 3x_2^2$, with discriminant -23.

4. Completing the square; Hermite's method of reduction

The elementary device of completing the square is crude but effective for many purposes. It is applicable to any form with $a_{11} \neq 0$, and consists in writing

$$4a_{11} f = (2a_{11} x_1 + a_{12} x_2 + \ldots + a_{1n} x_n)^2 + g(x_2, \ldots, x_n). \quad (38)$$

The form f is said to be HERMITE-REDUCED if it satisfies

$$\min f = |a_{11}|, \quad (39)$$

and also, except in the trivial case $n = 1$,

$$|a_{1j}| \leqslant |a_{11}|, \quad \text{for} \quad j = 2, \ldots, n, \quad (40)$$

and g defined by (38) is reduced. (41)

THEOREM 9. *Every form is equivalent to at least one Hermite-reduced form.*

Proof. We begin by transforming f, as explained in §1, so that (39) holds. If $n = 1$ this is the only condition; so we assume $n \geqslant 2$ and use induction on n.

By a suitable equivalence transformation with matrix of the shape $[1, T]$ we leave a_{11} unaltered and replace g by g^T, that is, choosing T suitably, by any $g' \sim g$. Applying the inductive hypothesis to the $(n-1)$-ary form g, we can take g' to be Hermite-reduced; this of course is trivial for $n = 2$. As in §1, we may keep the same notation and suppose that f satisfies (39) and (41). Now the parallel transformation (30) does not alter either a_{11} or g, and by suitable choice of its coefficients it takes f into a form satisfying (40) also, which completes the proof.

The discriminant of the form $4a_{11}f$ is clearly $(4a_{11})^n d(f)$, and its signature $(\operatorname{sgn} a_{11}) s(f)$. The discriminant and signature of the right member of (38) are easily found on taking it, by an obvious rational transformation with determinant $2a_{11}$, into $y_1^2 + g$. So using (9) and the rational invariance of the signature we find

$$d(g) = (-1)^{n-1} (16)^{[\frac{1}{2}n - \frac{1}{2}]} a_{11}^{n-2} d(f), \tag{42}$$

$$s(g) = (\operatorname{sgn} a_{11}) s(f) - 1. \tag{43}$$

In particular, g is positive if and only if f is definite, in which case $4a_{11}f$ is necessarily positive.

It is easy to see from (38) that for a definite form f any inequality of the shape $|f| \leqslant c$ has only a finite number of solutions, depending on c. For it implies

$$|2a_{11}x_1 + a_{12}x_2 + \ldots| \leqslant 2|a_{11}c|^{\frac{1}{2}} \quad \text{and} \quad g(x_2, \ldots) \leqslant 4|a_{11}c|;$$

and an obvious induction on n gives what is wanted. Applying this with $c = |a_{11}| - 1$, we see that by a finite calculation we can find the minimum of any given definite form, and transform it into one satisfying (39); thus f can be reduced in finitely many steps. To do the same for indefinite f, we may assume (40), (41), and $a_{11} \neq 0$; and we need a method which either verifies (39) or yields a solution of $0 < |f(\mathbf{x})| < |a_{11}|$. This seems very difficult.

This weakness can be avoided if we modify the definition, replacing condition (39) by

$$a_{11} \neq 0; \tag{44}$$

$$|a_{22}| \geqslant |a_{11}|, \quad \text{if} \quad n \geqslant 2 \text{ and } a_{22} \neq 0; \tag{45}$$

$$|a_{12}| = |a_{11}|, \quad \text{if} \quad n \geqslant 2 \text{ and } a_{22} = 0. \tag{46}$$

This of course modifies (41) also. We show that these conditions are all implied by (39), (40) and (41). This is clear as far as (44) and (45) are concerned, by the definition of the minimum. If $a_{22} = 0$, then the leading coefficient of g is $-a_{12}^2$, $\neq 0$ by (44) and (41). So if (46) is false, (40) gives $0 < |a_{12}| < |a_{11}|$. Now one of $|f(1, \pm 1, 0, \ldots, 0)|$ is $|a_{11}| - |a_{12}|$, > 0, $< |a_{11}|$, contradicting (39).

It is clear that if we replace the reduction conditions (39)–(41) by (40)–(46) Theorem 9 remains valid, and it becomes possible to transform any given form into an equivalent reduced one in a finite number of steps.

5. Crude estimate for min f; classes with given n, d

We prove a theorem which is very crude but has an important application.

THEOREM 10. *For every form f we have*

$$\frac{4^{-[\frac{1}{2}(n-1)^2]} \min^n f}{|d(f)|} \leqslant \left\{ \begin{array}{ll} 1, & \text{always,} \\ 3^{-\frac{1}{2}n(n-1)}, & \text{if } f \text{ is not a zero form.} \end{array} \right\} \tag{47}$$

Proof. By Theorem 9 and the invariance of the minimum under equivalence, it suffices to prove that

$$\frac{4^{-[\frac{1}{2}(n-1)^2]} |a_{11}|^n}{|d(f)|} \leqslant \left\{ \begin{array}{ll} 1, & \text{always,} \\ 3^{-\frac{1}{2}n(n-1)}, & \text{if } f \text{ is not a zero form,} \end{array} \right\} \tag{48}$$

when f is Hermite-reduced. That is, it is enough to deduce (48) from (39)–(41). It is also sufficient to deduce (48) from (40)–(46); for (44) implies $\min f \leqslant |a_{11}|$, and we have seen that (39)–(41) imply (44)–(46). (48) is trivial for $n = 1$, since then $d = a_{11}$; so we assume $n \geqslant 2$ and use induction.

If $a_{22} \neq 0$, (40) and (45) give $|b_{22}| \geqslant 3a_{11}^2$, where

$$b_{22} = 4a_{11}a_{22} - a_{12}^2$$

is the leading coefficient of g. If $a_{22} = 0$, (46) gives $|b_{22}| = a_{12}^2 = a_{11}^2$. So we have

$$|b_{22}| \geqslant \left\{ \begin{array}{l} a_{11}^2, \quad \text{always,} \\ 3a_{11}^2, \quad \text{if } f \text{ is not a zero form.} \end{array} \right\} \qquad (49)$$

The inductive hypothesis, that is, (48) with g, $n-1$, b_{22} for f, n, a_{11}, can be put, with a little calculation, using (42), into the shape

$$\frac{4^{-[\frac{1}{2}(n-1)^2]} |b_{22}|^{n-1}}{|a_{11}^{n-2} d(f)|} \leqslant \left\{ \begin{array}{l} 1, \quad \text{always,} \\ 3^{-\frac{1}{2}(n-1)(n-2)}, \quad \text{if } g \text{ is not a zero form.} \end{array} \right.$$

To deduce (48) from this inequality and (49) we have only to show that if g represents zero then f does. But this is clear; for (38) and $g(x_2, \ldots) = 0$ give

$$f(-a_{12} x_2 - \ldots - a_{1n} x_n, \ 2a_{11} x_2, \ldots, 2a_{11} x_n) = 0.$$

THEOREM 11. *There are at most finitely many classes of forms with given rank n and given discriminant d.*

Proof. By Theorem 9, it suffices to show that there are at most finitely many Hermite-reduced forms with given n, d. This being trivial for $n = 1$, we use induction. By the inductive hypothesis and (38), it is sufficient to show that there are at most finitely many possibilities for a_{11}, a_{12}, ..., a_{1n}, $d(g)$ when f is Hermite-reduced with given n, d. Now Theorem 10 gives what is wanted as far as a_{11} is concerned; and then (40) and (42) give bounds for a_{12}, ..., a_{1n} and $d(g)$. This completes the proof.

It follows *a fortiori* that there are at most finitely many classes with given n, s, d. Sometimes there is only one. As an example, suppose $n = s = 3$, $d = -4$. Then (48) gives $|a_{11}| \leqslant 1$, whence by (44) and $s = n$ we have $a_{11} = 1$. Now (42) gives $d(g) = -64$, so $b_{22} = 3$ or 4 by (48) and (49) and $s = n$. But with $d = -64$ and $a = 3$, (37) is insoluble, so g cannot represent 3, and we have $b_{22} = 4$. Now we find by Theorem 7 that g must be $4(x_2^2 + x_3^2)$. Using (40) we have $a_{12} = a_{13} = 0$, or f would not be integral. Thus we see that the form $x_1^2 + x_2^2 + x_3^2$ is the only reduced positive form with $n = 3$, $d = -4$. From this fact the classical three-square theorem, that this form represents properly every positive integer not congruent to 0, 4 or 7 modulo 8, can be deduced. But it is most simply obtained as a special case of a much more general result. (See Theorem 51.)

6. Zero forms

The results of the last section are numerically very weak for zero forms. They can be improved if we consider forms with $a_{11} = 0$; but then we need a substitute for the process of completing the square. For convenience later we begin with a result valid for all forms, whether representing zero or not.

THEOREM 12. *Every form with* $n \geqslant 3$ *is equivalent to one with the same leading coefficient and with*

$$a_{1j} = 0 \quad for \quad j = 3, \dots, n; \tag{50}$$

or more generally with

$$a_{ij} = 0 \quad for \quad |i-j| \geqslant 2. \tag{51}$$

Proof. We can write the given form as $a_{11}x_1^2 + hx_1\mathbf{c}'\mathbf{y} + \dots$, where (i) h is an integer, (ii) $\mathbf{c}'$ is a primitive row vector with $n-1$ elements, (iii) $\mathbf{y} = \{x_2, \dots, x_n\}$, and (iv) the terms not written do not involve x_1. An equivalence transformation with matrix $[1, T]$ takes this form into $a_{11}x_1^2 + hx_1\mathbf{c}'T\mathbf{y} + \dots$, which has the property (50) if $\mathbf{c}'T = (1, \mathbf{0}')$, or $T'^{-1}\{1, \mathbf{0}\} = \mathbf{c}$. By Theorem 1 we can find a T with this property.

To obtain (51), transform further, using equivalence transformations with matrices of the shape $[1, 1, T]$. These do not alter a_{11}, nor affect (50).

It is easily seen that for even n (51) implies $d \equiv (a_{12}a_{34}\dots)^2$ (mod 4). Since every square is congruent to 0 or 1 modulo 4, we see by Theorem 12 and the invariance of d under equivalence that

$$d \equiv 0 \text{ or } 1 \pmod 4 \text{ if } n \text{ is even.} \tag{52}$$

THEOREM 13. *Every zero form is equivalent to one of the shape*

$$x_2(a_{12}x_1 + a_{22}x_2 + a_{32}x_3 + \dots + a_{n2}x_n) + g(x_3, \dots, x_n) \tag{53}$$

if $n \geqslant 3$, *or* $a_{12}x_1x_2 + a_{22}x_2^2$ *if* $n = 2$, *with*

$$|a_{i2}| \leqslant \tfrac{1}{2}a_{12} \quad for \quad i = 2, \dots, n, \tag{54}$$

and $\quad\quad a_{i2} = 0 \quad for \quad i = 3, \dots, n \text{ if } a_{22} = 0. \tag{55}$

Proof. By Theorems 1 and 12 we may assume $a_{11} = 0$, and (50), and then the form is of the shape (53). Now $d \neq 0$ clearly implies $a_{12} \neq 0$, and so by putting if necessary $-x_1$ for x_1 we may suppose

a_{12} positive. Now the parallel transformation (30) does not alter a_{12} or g, and it replaces the coefficient a_{i2}, $i \geqslant 2$, by $a_{i2} + h_i a_{12}$. Choosing suitable integers h_i, these new coefficients satisfy (54).

To obtain (55), note that the a_{12} in (53) is the h of the proof of Theorem 12; it is the g.c.d. of the coefficients of $x_1 x_2, \ldots, x_1 x_n$ in the form with $a_{11} = 0$ to which we applied Theorem 12. If $a_{22} = 0$, we can interchange x_1, x_2 and apply Theorem 12 again, with h' in place of h, where $h' = $ g.c.d.$(a_{12}, a_{32}, \ldots, a_{n2})$ is numerically smaller than h if (54) holds and (55) fails. Assuming as we may that the form is not equivalent to one of the same shape (53) but with smaller a_{12}, we obtain a contradiction which completes the proof.

It is clear that a unary form (with $d = a_{11} \neq 0$) can never represent zero. A binary form represents zero if and only if d is a perfect square; the 'if' follows from $f(\pm d^{\frac{1}{2}} - a_{12}, 2a_{11}) = 0$, and the 'only if' from $d = a_{12}^2$ for $a_{11} = 0$. We now solve the crucial problem of finding a necessary and sufficient condition for a ternary form to represent zero.

THEOREM 14. *A ternary form represents zero if and only if it represents properly a multiple of* $d \prod\limits_{p|d} p$.

Proof. The 'only if' is trivial. The sufficiency of the condition will be proved first for the case $d = \pm 1$, in which it is vacuous. If $n = 3$ and f is not a zero form, then from (47), with $|d| = 1$, we find $\min f < 1$ since $3^3 > 4^2$. This is a contradiction, since $\min f$ is a positive integer by definition.

By hypothesis, and Theorems 1, 12, we may suppose

$$\left. \begin{array}{l} f \equiv x_2(a_{12}x_1 + a_{22}x_2 + a_{32}x_3) + a_{33}x_3^2 \\ d \equiv a_{12}^2 a_{33} \end{array} \right\} \pmod{d \prod_{p|d} p}. \qquad (56)$$

Write $\qquad d_1 = $ g.c.d. $(a_{12}, d), \quad d_2 = $ g.c.d. $(a_{33}, d);$

then $(56)_2$ gives $d_1^2 d_2 = |d|$. The form

$$f' = d_2^{-1} f(d_1^{-1} x_1, d_2 x_2, x_3)$$

is integral by $(56)_1$. Since $d(f') = d_2^{-3} d_1^{-2} d_2^2 d = \pm 1$, f' is a zero form by what we have just proved. It follows that

$$f(d_1^{-1} x_1, d_2 x_2, x_3)$$

is a zero form; and as the property of representing zero is rationally invariant, the theorem follows.

We state here the corresponding results for $n \geqslant 4$, which we express differently later (see Theorem 22) and prove by other methods.

THEOREM 15. (i) *If* $n = 4$, *then* f *represents zero if and only if it is indefinite and represents properly a multiple of* d.

(ii) *If* $n \geqslant 5$, *then* f *represents zero if and only if it is indefinite.*

7. Representation of zero subject to congruence conditions

A binary form, if it represents zero at all, does so essentially in only two ways; for $f(x_1, x_2) = 0$ holds for just two values of the ratio x_1/x_2. But a zero form with $n \geqslant 3$ represents zero in infinitely many ways, as is clear from (53). We investigate here the possibility of choosing a solution of $f(\mathbf{x}) = 0$ satisfying a congruence condition on $\mathbf{x}$. As we do not apply the result till later, and the proof is a little tedious, it may be worth while to explain the object of it. We are interested in Diophantine equations $f(\mathbf{x}) = a$, with $n(f) \geqslant 2$ as the case $n(f) = 1$ is trivial. We try to reduce the general problem to the case $a = 0$, since this case is easier and we have already made some progress with it. We may approach the general case gradually by first solving $f(\mathbf{x}) - ax_{n+1}^2 = 0$, and then trying to impose conditions on x_{n+1}. Ultimately we want to have $x_{n+1} = \pm 1$; but it will be useful if we can first make x_{n+1} prime to any given non-zero integer. And we may need to make $\mathbf{x}$ satisfy some congruence condition. Such results will be deduced later from:

THEOREM 16. *Suppose that* $n \geqslant 3$, f *is a zero form, and that for some fixed* $\mathbf{t}$, q ($q \neq 0$) *and every* $m \neq 0$ *the congruences*

$$f(\mathbf{y}) \equiv 0 \; (\mathrm{mod}\, m), \quad \mathbf{y} \equiv \mathbf{t} \; (\mathrm{mod}\, q) \tag{57}$$

are simultaneously soluble (with integral $\mathbf{y}$). *Then there exist an integer* h *and an integral vector* $\mathbf{x}$ *with*

$$f(\mathbf{x}) = 0, \quad \mathbf{x} \equiv h\mathbf{t} \; (\mathrm{mod}\, q), \quad \text{g.c.d.} \; (h, q) = 1. \tag{58}$$

Proof. Since (57) and (58) are unaltered on replacing f by an equivalent form f^T, and $\mathbf{x}$, $\mathbf{y}$, $\mathbf{t}$ by $T^{-1}\mathbf{x}$, $T^{-1}\mathbf{y}$, $T^{-1}\mathbf{t}$, we may by

Theorem 13 assume f to be a form of the shape (53). It will now suffice to find integral $\mathbf{x}$, h, k ($hk \neq 0$) such that

$$x_2(a_{12}x_1 + a_{22}x_2 + \ldots + a_{n2}x_n) + g(x_3, \ldots, x_n) \equiv 0 \pmod{a_{12}qk},$$

$$\mathbf{x} \equiv h\mathbf{t} \pmod{q}, \quad x_2 = k, \quad \text{g.c.d. } (h, a_{12}qk) = 1. \quad (59)$$

For a solution of (59) yields one of (58) on adding a suitable integral multiple of q to x_1.

In place of (59), it will suffice to solve (with integral $\mathbf{z}, h', k$)

$$z_2(a_{12}z_1 + \ldots) + g(z_3, \ldots) \equiv 0 \pmod{a_{12}qk},$$

$$\mathbf{z} \equiv \mathbf{t} \pmod{q}, \quad z_2 = h'k, \quad \text{g.c.d. } (h', a_{12}qk) = 1. \quad (60)$$

For if (60) holds, then (59) can be satisfied by choosing $\mathbf{x}$ so that $\mathbf{x} \equiv h\mathbf{z} \pmod{a_{12}qk}$, $x_2 = k$, these conditions being consistent if we choose h such that $hh' \equiv 1 \pmod{a_{12}qk}$.

We shall take k to be an integer whose prime factors all divide $a_{12}q$. Defining u, v, w $= u(p), v(p), w(p)$ by

$$p^u \| a_{12}, \quad p^v \| q, \quad p^w \| k,$$

we see that it is enough to show that for each $p \mid a_{12}q$ we can solve, with integral $\mathbf{z}$,

$$z_2(a_{12}z_1 + \ldots) + g(z_3, \ldots) \equiv 0 \pmod{p^{u+v+w}},$$

$$\mathbf{z} \equiv \mathbf{t} \pmod{p^v}, \quad p^w \| z_2, \quad (61)$$

for some $w = w(p) \geqslant 0$.

Now the hypothesis (57), with $m = p^{3u+3v}$, shows that there exists an integral $\mathbf{y}$ with

$$y_2(a_{12}y_1 + \ldots) + g(y_3, \ldots) \equiv 0 \pmod{p^{3u+3v}}, \quad \mathbf{y} \equiv \mathbf{t} \pmod{p^v}. \quad (62)$$

We may clearly suppose that $\mathbf{z} = \mathbf{y}$ does not (for any $w < 2u + 2v$) satisfy (61); and this gives us $p^{2u+2v} \mid y_2$, whence

$$p^v \mid t_2, \quad p^{2u+2v} \mid g(y_3, \ldots). \quad (63)$$

Without disturbing (62) and (63) we may add suitable multiples of p^{3u+3v} to the y_i, and so suppose that $g(y_3, \ldots) \neq 0$. (This fails only if g is identically 0, making $d = 0$.) We may therefore write

$$g(y_3, \ldots) = ap^{2u+2v+r}, \quad r \geqslant 0, \quad p \nmid a; \quad a_{32}y_3 + \ldots = b.$$

We now put $z_i = y_i$ for $i = 3, \ldots, n$;

$$z_2 = a_{12} p^v z_2' \quad \text{and} \quad z_1 = p^v z_1' - a_{22} p^v z_2' + t_1.$$

Because of $(63)_1$ this satisfies $(61)_2$, and (61) reduces to

$$a_{12} p^v z_2'(a_{12} p^v z_1' + a_{12} t_1 + b) \equiv -p^{2u+2v+r} a \pmod{p^{u+v+w}},$$

$$p^{w-u-v} \| z_2'.$$

Since $p \nmid a$, these two conditions hold for some $w \leqslant 2u + 2v + r$ if we choose integers z_1', z_2' so that

$$a_{12} p^v z_2'(a_{12} p^v z_1' + a_{12} t_1 + b) \equiv -p^{2u+2v+r} a \pmod{p^{3u+3v+2r+1}}.$$

This congruence is soluble for z_2' if we choose z_1' so that

$$p^{2u+2v+r+1} \nmid a_{12} p^v (a_{12} p^v z_1' + a_{12} t_1 + b).$$

And if this fails for $z_1' = 0$ and for $z_1' = 1$, then since $r \geqslant 0$ we have $p^{2u+1} \mid a_{12}^2$, contrary to the definition of u. This contradiction completes the proof.

8. Reciprocal forms

In the rest of this chapter we shall develop the theory of reduction and estimation of the minimum a little further. But as we have already obtained all the key results (Theorems 11, 14 and 16) that we need later, we shall omit or condense many of the proofs. The inductive argument of Theorem 10 does not give the best estimates for $\min f$ in terms of $|d|^{1/n}$ without complicated calculations; so we describe an alternative argument which sometimes does.

Two forms, with matrices say A, B, are said to be RECIPROCAL if $AB \,(= BA)$ is a scalar multiple of the identity matrix. If so, then each is rationally related to a multiple of the other; for $B'AB = BAB = aB$ if $AB = aI$. It follows that if either of them is a zero form the other is, and that their signatures are equal in absolute value. In particular, every form f is reciprocal to its ADJOINT FORM, which we denote by $\mathrm{adj} f$ and define by

$$A(\mathrm{adj} f) = \begin{cases} (-1)^{\frac{1}{2}n-1} \mathrm{adj}\, A(f) & (n \text{ even}), \\ 2(-1)^{\frac{1}{2}n-\frac{1}{2}} \mathrm{adj}\, A(f) & (n \text{ odd}). \end{cases}$$

This definition is so constructed that $\mathrm{adj}\,f$ is (with f) an integral form, whose leading coefficient is the discriminant of the $(n-1)$-ary form $f(0, x_2, ..., x_n)$. It gives

$$d(\mathrm{adj}\,f) = \begin{cases} d^{n-1}(f) & (n \text{ even}), \\ 4^{n-1}d^{n-1}(f) & (n \text{ odd}). \end{cases} \tag{64}$$

Now note that equivalent forms have equivalent reciprocals (and equivalent adjoints): $(T'AT)(T_1'BT_1) = aI$ if $AB = aI$ and $T_1 = T'^{-1}$. We may therefore assume, in estimating the minimum of f, that

$$\min \mathrm{adj}\,f = |d(f_1)|, \quad \text{where} \quad f_1 = f(0, x_2, ..., x_n). \tag{65}$$

We seek the best possible estimate for the function

$$|d(f)|^{-1} \min^n f, \quad \text{say } \mu(f).$$

Clearly we have

$$\min f \leqslant \min f_1. \tag{66}$$

Now by (64)–(66) we have

$$\begin{aligned} \mu^{n-1}(f) &= |d(f)|^{1-n} \min^{n(n-1)} f \\ &\leqslant |d(f)|^{1-n} \min^{n(n-1)} f_1 \\ &= |d(f)|^{1-n} \mu^n(f_1) |d(f_1)|^n \\ &= |d(f)|^{1-n} \mu^n(f_1) \min^n \mathrm{adj}\,f \\ &= |d(f)|^{1-n} \mu^n(f_1) |d(\mathrm{adj}\,f)| \,\mu(\mathrm{adj}\,f), \end{aligned}$$

whence

$$\frac{\mu^{n-1}(f)}{\mu(\mathrm{adj}\,f)\,\mu^n(f_1)} \leqslant \begin{cases} 1 & (n \text{ even}), \\ 4^{n-1} & (n \text{ odd}). \end{cases}$$

There is clearly no loss of generality in considering only forms with

$$\mu(f) \geqslant \mu(\mathrm{adj}\,f);$$

and for these we deduce

$$\mu^{n-2}(f) \leqslant \begin{cases} \mu^n(f_1) & (n \text{ even}), \\ 4^{n-1}\mu^n(f_1) & (n \text{ odd}). \end{cases} \tag{67}$$

We mention three applications. First suppose $n(f) = 4$, $s(f) = 0$, and f not a zero form. Then it is easy to see that $s(f_1) = \pm 1$ and f_1 is not a zero form. From a known result we have $\mu(f_1) \leqslant \frac{1}{6}$, and we deduce $\mu(f) \leqslant \frac{1}{36}$. This is best possible, as shown by the example

$$f = x_1^2 + x_1 x_2 + x_2^2 - 2(x_3^2 + x_3 x_4 + x_4^2),$$

with discriminant 36 and minimum 1.

Next let $n(f) = s(f) = 4$. Then f_1 is a positive ternary form, and it is known that $\mu(f_1) \leqslant \frac{1}{2}$, which gives $\mu(f) \leqslant \frac{1}{4}$. This is again best possible, as shown by

$$f = \sum_{i=1}^{3} (x_i + \tfrac{1}{2}x_4)^2 + \tfrac{1}{4}x_4^2, \quad d(f) = 4, \quad \min f = 1.$$

Thirdly, the second of the two results $\mu(f) \leqslant 1$ for $n = s = 7$, 8 follows by (67) from the first, and is best possible because of the example

$$f = \frac{1}{4} \sum_{i=1}^{4} (2x_i + l - x_{i+4})^2 + \tfrac{1}{4}(x_5^2 + x_6^2 + x_7^2 + x_8^2),$$

$$l = x_5 + x_6 + x_7 + x_8, \quad d(f) = 1, \quad \min f = 1.$$

9. Minkowski reduction of positive forms

Till the end of this chapter we shall consider positive forms only, and shall allow the coefficients to take *real* values. This enables us to use elementary analysis; but the final results and some of the proofs involve integral forms only. We begin with a definition which agrees with that of §2 for the case $n = 2$.

A positive form f is said to be MINKOWSKI-REDUCED if its coefficients satisfy the inequalities

$$f(\mathbf{x}) \geqslant a_{ii} \quad \text{whenever} \quad \text{g.c.d. } (x_i, ..., x_n) = 1, \tag{68}$$

$$a_{i,i+1} \geqslant 0 \quad \text{for} \quad i = 1, ..., n-1. \tag{69}$$

It may be noted that (68) implies $\min f = a_{11}$ and

$$a_{11} \leqslant a_{22} \leqslant ... \leqslant a_{nn}. \tag{70}$$

It is to be understood that $\mathbf{x}$ in (68) ranges over all primitive vectors, and that g.c.d. $(x_i, ..., x_n) = 1$ means $x_n = \pm 1$ in case $i = n$.

To prove that every positive form is equivalent to a Minkowski-reduced form, we need only consider the inequalities (68). For the conditions (69) can be satisfied, without altering the a_{ii}, by changing the signs of one or more variables if necessary. We show that if (68) fails we can transform f into an equivalent form whose coefficients b_{ij} satisfy $b_{ii} < a_{ii}$, $b_{jj} = a_{jj}$ for $j < i$ (if $i > 1$). Regarding any such form as simpler than f, the result follows on noting that there cannot be an infinite sequence of equivalent forms, each simpler than its predecessor. (This follows from the fact noted in §4 that any inequality $f(\mathbf{x}) \leqslant c$ has only finitely many solutions; whence in particular $\min f$ defined in §1 for integral f exists also for real positive f.)

To effect the desired simplification we need an equivalence transformation whose matrix T has its jth column vector equal to that of the identity for $j = 1, ..., i-1$ in case $i > 1$, while its ith column vector is equal to an $\mathbf{x}$ for which (68) fails. This matrix can be constructed as in Theorem 1.

If we take $\mathbf{x}$ in (68) to be the sum or difference of two base vectors we see that a Minkowski-reduced form must satisfy

$$|a_{ij}| \leqslant a_{ii} \quad \text{for} \quad 1 \leqslant i < j \leqslant n. \tag{71}$$

Taking $\mathbf{x} = \{1, -1, 1, 0, ..., 0\}$ in case $n \geqslant 3$, we see that we must also have
$$a_{12} - a_{13} + a_{23} \leqslant a_{11} + a_{22}. \tag{72}$$

In case $n = 2$ we have already seen that conditions (70) and (71) imply all others of the infinite system of inequalities (68), and also, with $a_{11} > 0$, imply that f is positive. This ceases to be true for $n \geqslant 3$; but for $n = 3$ only one other inequality, namely (72), is needed. We leave this to the reader, but sketch the proof of

THEOREM 17. *A positive ternary Minkowski-reduced form satisfies*
$$|d| \geqslant 4a_{11}a_{22}a_{33} - a_{11}a_{22}^2 - a_{11}^2 a_{33}.$$

Proof. Expanding $|d|$, $= -d$, we write the inequality to be proved as

$$a_{11}a_{22}^2 + a_{11}^2 a_{33} + a_{12}a_{13}a_{23} - a_{11}a_{23}^2 - a_{22}a_{13}^2 - a_{33}a_{12}^2 \geqslant 0.$$

The left member is a quadratic polynomial in a_{12}, with negative leading coefficient; hence when all the other a_{ij} are fixed it takes its least value for a_{12} at one end of the interval to which (69), (71) and (72) restrict it. Hence we may suppose that $a_{12} = 0$, $a_{12} = a_{11}$, or that equality holds in (72). There is a similar argument with the suffixes 1, 3 or 2, 3 interchanged.

If equality holds in (72), we may, without disturbing (72), vary $a_{12} + a_{23}$ while keeping fixed $a_{12} - a_{23}$, a_{13} and all the a_{ii}. This shows that we may assume equality in one of (69) or (71). Again we may permute the suffixes; and finally we see that we may assume three cases of inequality to hold in (69), (71) and (72) together. If one of these occurs in (69) the desired result is easy; if not, two of them must occur in (71), and the argument becomes straightforward.

We deduce from Theorem 17 the well-known inequalities

$$|d| \geqslant 2a_{11}a_{22}a_{33} \geqslant 2a_{11}^3 \geqslant 2\min^3 f.$$

Here by Theorem 17 equality requires $a_{11} = a_{22} = a_{33}$; but on reworking the foregoing argument we see easily that it requires also either all three of $a_{12} = 0$ or a_{11}, $a_{23} = 0$ or a_{11}, $a_{13} = \pm a_{11}$ or two of these and equality in (72), implying the third or $a_{13} = 0$.

Now we see that $|d| = 2\min^3 f$ implies that f is equivalent to a multiple of the form

$$x_1^2 + x_2^2 + x_3^2 + x_2 x_3 + x_3 x_1 + x_1 x_2,$$

or of $x_1^2 + x_2^2 + x_3^2 + x_1(x_2 + x_3)$. But by putting $\{x_1 + x_2, -x_2, x_3\}$ for $\mathbf{x}$ in the former we see that these two cases are equivalent.

10. Extreme forms

In this and the next section, we regard a quadratic form f as a point, with the a_{ij} as co-ordinates, in $\frac{1}{2}n(n+1)$-dimensional space. We regard n as fixed, and wish to find the upper bound of the function $\mu(f) = \min^n f / |d(f)|$; we know by Theorem 10 that such a bound exists. We sketch the proof of

THEOREM 18. (i) *The region of space in which f is a positive reduced form, in either Hermite's or Minkowski's sense, is defined by the inequality $a_{11} > 0$ together with a finite set of inequalities each of which asserts that some continuous function of the a_{ij} is non-negative.*

(ii) *The region in which f is positive and reduced (in either sense) and satisfies $a_{11} = 1$ and $|d| \leqslant 2^n$ is closed and bounded, and in it $\mu(f)$ has the same upper bound as for all positive f.*

(iii) *$\mu(f)$ attains its upper bound.*

Proof. (i) is clear, except for the finiteness, from (39) to (41), (68) and (69). For Hermite reduction the finiteness is quite easy to prove (by induction); for Minkowski reduction it is difficult.

(ii) We get rid of the open inequality $a_{11} > 0$ by putting $a_{11} = 1$; this does not alter the range of values of $\mu(f)$, since it is easy to see that $\mu(a_{11}^{-1}f) = \mu(f)$. We are now working in a region in which $\min f = 1$ and so $\mu(f) = |d|^{-1}$; thus we seek the lower bound of $|d|$ in this region. The lower bound is not altered by the restriction $|d| \leqslant 2^n$; for by taking $f = \mathbf{x}'\mathbf{x}$ we see that $|d|$ takes values satisfying this condition. Now we have a region all of whose defining inequalities are of the type described in (i), and we need only show that it is bounded. This follows easily from (70), $a_{11} = 1$, and $|d| \leqslant 2^n$ if we deduce from the reduction conditions an upper bound for $|d|^{-1}a_{11}...a_{nn}$. For Hermite reduction this is straightforward; indeed, it is not difficult to show that $(48)_2$ holds with $a_{11}...a_{nn}$ in place of a_{11}^n. For Minkowski reduction the proof is again more difficult.

(iii) Using either of the two cases of (ii), we have only to note that $|d|$, which is a continuous function of the a_{ij}, attains its lower bound in the closed and bounded region.

A form f for which $\mu(f)$ attains its upper bound is said to be ABSOLUTELY EXTREME; one with the weaker property that any sufficiently small variation of the coefficients leads to a form f' with $\mu(f') \leqslant \mu(f)$ is called EXTREME.

11. Perfect forms

The integral values of $\mathbf{x}$ for which $f(\mathbf{x}) = \min f$ may be called the MINIMAL VECTORS of f, and denoted by $\mathbf{m}_1, \mathbf{m}_2, ...$. The form f is said to be PERFECT if the equations

$$f(\mathbf{m}_1) = \min f, \quad f(\mathbf{m}_2) = \min f, ...$$

determine the a_{ij} uniquely (in terms of $\min f$ and $\mathbf{m}_1, \mathbf{m}_2, ...$).

THEOREM 19. *Every extreme form is perfect.*

Proof. Suppose f is not perfect; we shall prove that it is not extreme. By the definition just given, the linear equations $f(\mathbf{m}_1) = \min f, ...$ do not determine the a_{ij} uniquely. There is therefore a form, say g, which may have discriminant 0 but does not vanish identically, such that $g(\mathbf{m}_1) = g(\mathbf{m}_2) = ... = 0$. It follows that for all θ the form $f(\mathbf{x}) + \theta g(\mathbf{x})$ is equal to $\min f$ whenever $\mathbf{x}$ is one of the $\mathbf{m}_i$. It is easily proved that if $|\theta|$ is small enough we have $f(\mathbf{x}) + \theta g(\mathbf{x}) > \min f$ whenever $\mathbf{x}$ is not one of the $\mathbf{m}_i$; and this makes $\min (f(\mathbf{x}) + \theta g(\mathbf{x})) = \min f$.

The result will now follow if, writing f' for $f(\mathbf{x}) + \theta g(\mathbf{x})$, we prove that for some small θ we have $|d(f')| < |d(f)|$, implying, with $\min f' = \min f, \mu(f') < \mu(f)$. That is, A, B being the matrices of f, g, we have to prove that $|A + \theta B|$ is numerically less than $|A|$ for some arbitrarily small θ.

In proving this, we do not use the hypothesis that f is not perfect, but only that f is positive and A, B are symmetric and $B \neq O$. Hence we may suppose (Theorem 3) that A is diagonal, with positive elements $2a_{ii}$. If R is the real diagonal matrix with elements $(2a_{ii})^{-\frac{1}{2}}$ and $R'BR = C$, then C is symmetric and $\neq O$, and the problem reduces to proving that

$$|I + \theta C| < 1,$$

for some sufficiently small θ of suitable sign.

A straightforward calculation now shows that the value of $|I + \theta C|$ is

$$(1 + \tfrac{1}{2}\theta \sum_i c_{ii})^2 - \tfrac{1}{2}\theta^2 \sum_i \sum_j c_{ij}^2 + \ldots,$$

omitting terms in θ^3 and higher powers. The result follows.

It is not difficult to find a k_n depending only on n and such that all the elements of the minimal vectors of every reduced form (whether perfect or not) lie between $\pm k_n$. Hence we have in principle a method of calculating the upper bound of $\mu(f)$ for positive forms of any given rank n.

It follows easily from the definition that a perfect form is never disjoint, and is always a multiple of an integral form.

CHAPTER 3

THE RATIONAL INVARIANTS

1. p-adic squares; the Legendre symbol

We study the congruence properties, in relation to a given prime p, of a unary form $a_{11}x_1^2$. It is sufficient to consider the case $a_{11} = 1$, and to determine the integers $a \neq 0$ such that the congruence

$$x_1^2 \equiv a \pmod{p^r} \tag{73}$$

is soluble for every r; we call such integers a p-ADIC SQUARES. If $p \mid a$, then (73) is soluble, if at all, only with $p \mid x_1$, $p^2 \mid x_1^2$. Thus a is never a p-adic square if $p \parallel a$, and is a p-adic square if and only if $p^{-2}a$ is one in case $p^2 \mid a$. If $p \nmid h$, then a is a p-adic square if and only if ah^2 is one. The 'only if' is trivial; and for the 'if', note that a solution of $y_1^2 \equiv ah^2 \pmod{p^r}$ yields one of (73) on solving (for x_1) the trivial linear congruence $hx_1 \equiv y_1 \pmod{p^r}$. By what we have said about the case $p \mid a$, this result remains valid if the condition $p \nmid h$ is weakened to $h \neq 0$.

It will suffice to determine the p-adic squares not divisible by p; and we show first that with $p \nmid a$ it suffices to take $r = 3$ or 1 in (73) according as $p = 2$ or $p > 2$. Take first $p = 2$, and using induction on r suppose that $r \geq 4$ and $x_1 = u$ satisfies (73) with $r - 1$ in place of r. We show that either $x_1 = u$ or $x_1 = u + 2^{r-2}$ satisfies (73). This is clear since u must be odd, and so

$$(u + 2^{r-2})^2 = u^2 + 2^{r-1}u + 2^{2r-4} \equiv u^2 + 2^{r-1} \pmod{2^r}.$$

For odd p, suppose $r \geq 2$, $u^2 \equiv a \pmod{p^{r-1}}$, whence $p \nmid a$ gives $p \nmid u$. Now (73) can be solved by putting $x_1 = u + p^{r-1}v$. For we have

$$(u + p^{r-1}v)^2 \equiv u^2 + 2up^{r-1}v \pmod{p^r},$$

and so (73) reduces to the linear congruence $2uv \equiv p^{1-r}(a - u^2) \pmod{p}$, which is soluble since $p \nmid 2u$.

We now see, since $(\pm 1)^2 \equiv (\pm 3)^2 \equiv 1 \pmod 8$, that an odd integer a is a 2-adic square if and only if $a \equiv 1 \pmod 8$. In case $p \nmid 2a$, a is called a QUADRATIC RESIDUE modulo p if a is a p-adic square; and we see that this is so if and only if $a \equiv x_1^2 \pmod p$ for

x_1 equal to one of $1, ..., \frac{1}{2}(p-1)$. Plainly there are $a \not\equiv 0 \pmod{p}$ for which this is not the case; any such a is called a QUADRATIC NON-RESIDUE modulo p ($\neq 2$). The LEGENDRE SYMBOL $(a|p)$ is defined to have the value 0 if $p|a$, 1 if a is a quadratic residue, and -1 otherwise. Hence it has the property $(a|p) = (b|p)$ if $a \equiv b \pmod{p}$. The integers x_1^2, $x_1 = 1, ..., \frac{1}{2}(p-1)$, are all incongruent modulo p; for $h^2 \equiv k^2 \pmod{p}$ implies $p|h-k$ or $p|h+k$. Hence there are just $\frac{1}{2}(p-1)$ quadratic residues, and consequently also $\frac{1}{2}(p-1)$ non-residues, between 1 and $p-1$. If b is any particular non-residue, then all the non-residues are just the integers congruent to bx_1^2, $x_1 = 1, ..., \frac{1}{2}(p-1)$, modulo p, since these numbers are $\frac{1}{2}(p-1)$ incongruent non-residues. From these remarks we deduce the multiplicative property

$$(ab|p) = (a|p)(b|p) \tag{74}$$

of the Legendre symbol.

We now arrange all the non-zero integers in p-ADIC (QUADRATIC) CLASSES, in such a way that a, b are in the same class if and only if ab is a p-adic square. It is possible to do so (cf. (16)) because this last condition expresses a reflexive, symmetric and transitive relation between a and b; for a^2 is a p-adic square for $a \neq 0$, $ab = ba$, and ac is a p-adic square if each of ab, bc is one. It will be clear that the p-adic class of an integer $p^u a'$, $p \nmid a'$, is determined by the parity of u and the Legendre symbol $(a'|p)$ (or the residue of a' modulo 8 if $p = 2$). Thus the number of p-adic classes is 8 for $p = 2$, 4 for $p \neq 2$.

2. p-adic zero forms

By a p-ADIC ZERO FORM is meant a form f such that for every r there is an integral $\mathbf{x}$ with

$$p \nmid \mathbf{x}, \quad f(\mathbf{x}) \equiv 0 \pmod{p^r}. \tag{75}$$

A zero form as previously defined may be called a *rational* zero form if there is any risk of confusion, and is clearly a p-adic zero form for every p.

The property of being a p-adic zero form is rationally invariant. To prove this, consider the form f^R, R rational. Choose an integer $q \neq 0$ so that qR is integral; and supposing f^R to be a p-adic zero form, choose $\mathbf{y}$, integral, so that $p \nmid \mathbf{y}$ and $p^t \mid f^R(\mathbf{y})$, with suitable t. Now $\mathbf{x} = p^{-u}qR\mathbf{y}$ is integral, and $p \nmid \mathbf{x}$, for some $u \leqslant v$, where $p^v \parallel |qR|$, so that v is independent of t, $\mathbf{y}$. This $\mathbf{x}$ satisfies (75) if we choose $t \geqslant r + 2v$.

Trivially, f is a p-adic zero form if and only if af is one, for any $a \neq 0$. It is also clear that if any 'section' of f (obtained by putting one or more of the variables equal to 0) is a p-adic zero form then so is f.

In investigating rationally invariant properties we may by Theorem 3 assume f to be diagonal; and then it may be convenient to use the single suffix notation $f = a_1 x_1^2 + \ldots + a_n x_n^2$. By a suitable substitution $x_i \to h_i^{-1} x_i$ we may suppose the a_i square-free. In considering any property which is invariant under removal of the divisor of an imprimitive form, we may suppose $a_1 = 1$; for we may take f into $a_1 x_1^2 + a_2 a_1^2 x_2^2 + \ldots$, and then remove the factor a_1. Besides the property of being a p-adic zero form, we shall be interested in two others which have both these kinds of invariance, namely, the property of representing an integer in each p-adic class, and (for even n only) the p-adic class of d.

THEOREM 20. (i) *When* $n = 2$, f *is a* p-*adic zero form if and only if* d *is a* p-*adic square, and if so* f *represents an integer in each* p-*adic class.*

(ii) *When* $n \geqslant 3$, f *is a* p-*adic zero form if and only if a binary* p-*adic zero form splits off rationally from* f; *and in case* $n = 3$ *this is so if and only if* f *represents an integer in the* p-*adic class of* d.

(iii) f *represents an integer in the* p-*adic class of* $a \neq 0$ *if and only if* $f - ax_{n+1}^2$ *is a* p-*adic zero form.*

Proof. (i) By the foregoing remarks we may suppose

$$f = x_1^2 + a_2 x_2^2,$$

with $a_2 = -\frac{1}{4}d$ square-free. Now (75) is easily seen to be soluble if at all, with $x_2 = 1$. Thus the 'if' and the 'only if' follow immediately. Now with $-a_2$ a p-adic square, prime to p since it is square-free, it is clear that f takes modulo p^r, for every r, the same residues as $x_1^2 - x_2^2$. Hence the remaining assertion need only be proved for $f = x_1^2 - x_2^2$. Thus it becomes obvious—still more so if we transform rationally into $x_1 x_2$.

(ii) The 'if' is trivial. To prove the 'only if' suppose (75) soluble; clearly there is a solution with primitive $\mathbf{x}$, and so we may suppose $p^r | a_{11}$, for any suitable r. Now as in the proof of Theorem 14 we may also suppose $a_{13} = \ldots = a_{1n} = 0$, and then we must have $p^u \| a_{12}$ with some u such that $p^{2u} | d$ (if we choose r so that $p^r \nmid d$). We use Theorem 2 to split off the binary form $f(x_1, x_2, 0, \ldots, 0)$, and we notice that its discriminant, say d_1, is a p-adic square if we choose r so that $r > 2u$. For $d_1 = a_{12}^2 - 4a_{11}a_{22}$ satisfies $p^{2u} \| d_1$ and $d_1 \equiv a_{12}^2 \pmod{4p^{2u+1}}$.

In case $n = 3$, note that if f is rationally related to $f_1 + a_{33}x_3^2$ then d is in the p-adic class of $a_{33}d(f_1)$. Thus $a_{33}d$ is a p-adic square if and only if $d(f_1)$ is one.

(iii) The case in which f is a p-adic zero form is trivial by (i) and (ii). The 'only if' is straightforward; for if f represents b, with ab a p-adic square, then we may by Theorem 2 suppose

$$f = bx_1^2 + \ldots$$

and so

$$f(x_1, 0, \ldots, 0) - ax_{n+1}^2$$

is a p-adic zero form by (i).

Now suppose that $f - ax_{n+1}^2$ is a p-adic zero form. Then (see (75)), there exists for every r a solution of $f(\mathbf{x}) \equiv ax_{n+1}^2 \pmod{p^r}$ with either $p \nmid \mathbf{x}$ or $p \nmid x_{n+1}$. If for any r we have a solution of this congruence with $p^{r-2} \nmid ax_{n+1}^2$, then it is clear by the arguments of §1 that $f(\mathbf{x})$ and ax_{n+1}^2, whence also $f(\mathbf{x})$ and a, are in the same p-adic class. And if this never happens, then $f(\mathbf{x}) \equiv 0 \pmod{p^{r-2}}$ has always a solution with $p \nmid \mathbf{x}$; and f is a zero form, but this case has been dealt with.

3. The Hilbert symbol

The HILBERT SYMBOL $(a, b)_p$ is defined, for every pair of non-zero integers a, b and every prime p, to have the value 1 if the form $ax_1^2 + bx_2^2$ represents a p-adic square, -1 if not. Hence trivially $(a, b)_p = (b, a)_p$; and $(a, b)_p = 1$ if a or b is a p-adic square, or, by Theorem 20 (i), if $-ab$ is a p-adic square, the discriminant of the binary form being $-4ab$.

An equivalent definition, by Theorem 20 (iii) with $a = 1$, is that $(a, b)_p = 1$ if the ternary form $ax_1^2 + bx_2^2 - x_3^2$ is a p-adic zero form, -1 if not. Since this ternary form goes by an obvious rational transformation into $-a(ax_1^2 - abx_2^2 - x_3^2)$, we see by the invariant properties mentioned in §2 that

$$(a, b)_p = (a, -ab)_p. \tag{76}$$

It is clear that the value of the symbol is not altered if we transform the form $ax_1^2 + bx_2^2$ rationally into another diagonal form; nor if we replace the coefficient a by aq, where $p \nmid q$ and q is a p-adic square. Hence we see that

$$(a, b)_p = (a', b)_p \quad \text{if} \quad aa' \text{ is a } p\text{-adic square}; \tag{77}$$

in particular, this holds if aa' is a square.

By using these two formulae, we see that it suffices to evaluate $(a, b)_p$ for $p^2 \nmid ab$. The case $p \| ab$ is disposed of by the following formulae:

$$(a, b)_p = \begin{cases} (a \mid p) & \text{if} \quad p \nmid 2a, p \| b, \\ 1 & \text{if} \quad p = 2 \| b \text{ and } a \text{ or } a + b \equiv 1 \ (\text{mod } 8), \\ -1, & \text{otherwise.} \end{cases} \tag{78}$$

To prove these formulae, note that $ax_1^2 + bx_2^2$ will represent a p-adic square, if at all, with $p \nmid \mathbf{x}$. Now when $p \nmid a$, $p \| b$, we have $p \| f = ax_1^2 + bx_2^2$ if $p \mid x_1$, $p \nmid x_2$, whence f is not a p-adic square. If $p \nmid x_1$, then $(f \mid p) = (a \mid p)$ for $p \neq 2$ and $f \equiv a$ or $a + b$ (mod 8) if $p = 2$.

When ab is odd it is clear that $(a, b)_2 = 1$ if $a, b, a + 4b$ or $4a + b$ is congruent to 1 modulo 8, that is, if a or b is congruent to 1 modulo 4. In the remaining case $2 \nmid \mathbf{x}$ gives $f \equiv -1$ or 2 (mod 4), so that f is not a 2-adic square. Thus we have

$$(a, b)_2 = \begin{cases} 1 & \text{for} \quad a \equiv 1 \ (\text{mod } 4), b \text{ odd,} \\ -1 & \text{for} \quad a \equiv b \equiv -1 \ (\text{mod } 4). \end{cases} \tag{79}$$

Finally we prove

$$(a, b)_p = 1 \quad \text{if} \quad p \nmid 2ab. \tag{80}$$

It suffices to prove the solubility of the congruence $ax_1^2 + bx_2^2 \equiv 1$ (mod p). Now if this congruence were insoluble, the $p + 1$ integers $ax_1^2, 1 - bx_2^2$, with x_1, x_2 each ranging from 0 to $\frac{1}{2}(p - 1)$, would all be incongruent modulo p; clearly this is impossible.

From these formulae, using (74), we deduce

$$(a, b)_p \, (a', b)_p = (aa', b)_p. \tag{81}$$

We use the Hilbert symbol to prove

THEOREM 21. *If $n \geqslant 4$ and f is not a p-adic zero form, then $n = 4$ and d is a p-adic square. If $n = 4$, d is a p-adic square, and f is rationally related to $a_1 x_1^2 + \ldots + a_4 x_4^2$, then f is a p-adic zero form if and only if $(-a_1 a_2, -a_1 a_3)_p = 1$.*

Proof. We begin by reducing the case $n \geqslant 5$ to the case $n = 4$. Supposing $f = a_1 x_1^2 + \ldots + a_n x_n^2$, $n \geqslant 5$, we can take f rationally into $a_1 x_1^2 + a_2 x_2^2 + a_3 x_3^2 + b_4 x_4^2 + \ldots$, with $b_4 = a_4 y_1^2 + a_5 y_2^2$, for any integers y_1, y_2 for which this form does not vanish. Now if f is not a p-adic zero form then neither is the 4-ary form $a_1 x_1^2 + \ldots + b_4 x_4^2$. So it suffices to choose y_1, y_2 so that the discriminant $16 a_1 a_2 a_3 b_4$ of this form is not a p-adic square.

A little consideration shows that if this cannot be done then $a_1 a_2 a_3 a_4$ and $a_1 a_2 a_3 a_5$ are p-adic squares, and the form $y_1^2 + y_2^2$ represents p-adic squares only. Assuming this, take $c = -3$ if $p = 2$, and choose any c with $(c \,|\, p) = -1$ if $p \neq 2$. Then $c(y_1^2 + y_2^2)$ does not represent a p-adic square. That is, $(c, c)_p = -1$, contradicting (79) or (80).

This disposes of the case $n \geqslant 5$, so we suppose

$$n = 4, \quad f = a_1 x_1^2 + \ldots + a_4 x_4^2.$$

First suppose that $d = 16 a_1 a_2 a_3 a_4$ is a p-adic square; thus $-a_4$ is in the p-adic class of $d(a_1 x_1^2 + a_2 x_2^2 + a_3 x_3^2) = -4 a_1 a_2 a_3$. By Theorem 20 (ii) and (iii), f is a p-adic zero form if and only if

$$a_1 x_1^2 + a_2 x_2^2 + a_3 x_3^2$$

is one. Taking this ternary form rationally into a multiple of $-a_1 a_2 x_1^2 - a_1 a_3 x_2^2 - x_3^2$ (by putting x_3, $a_1 x_1$, $a_1 x_2$ for x_1, x_2, x_3) we can appeal again to Theorem 20 (iii) and see that f is a p-adic zero form if and only if $-a_1 a_2 x_1^2 - a_1 a_3 x_2^2$ represents a p-adic square.

There remains the case $n = 4$, d not a p-adic square, in which we must prove that f is a p-adic zero form. Here we have $d = d_1 d_2$, where $d_1, d_2 = -4 a_1 a_2$, $-4 a_3 a_4$ are the discriminants of the binary forms $f_1, f_2 = a_1 x_1^2 + a_2 x_2^2$, $a_3 x_3^2 + a_4 x_4^2$. We may suppose that neither of f_1, f_2 is a p-adic zero form, that is, that neither of d_1, d_2 is a p-adic square.

Since none of d_1, d_2, $d = d_1 d_2$ is a p-adic square, we see from (76)–(80) that as a ranges over the p-adic classes each of the Hilbert symbols $(a, d_1)_p$, $(a, d_2)_p$, also their product $(a, d)_p$, takes both the values ± 1. Using (81), we see that the pair of symbols $(a, d_1)_p$, $(a, d_2)_p$ takes all four of the pairs of values ± 1, ± 1. That is, we can find $a \neq 0$ with

$$(a, d_1)_p = (a_1, d_1)_p, \quad (a, d_2)_p = (-a_3, d_2)_p,$$

whence
$$(aa_1, d_1)_p = (-aa_3, d_2)_p = 1.$$

Using (76) this gives

$$(aa_1, aa_2)_p = (-aa_3, -aa_4)_p = 1.$$

This means that the binary forms

$$aa_1 x_1^2 + aa_2 x_2^2 \quad \text{and} \quad -aa_3 x_3^2 - aa_4 x_4^2$$

both represent p-adic squares; whence it is clear that af and f are p-adic zero forms. This completes the proof; but it may be remarked that the construction works also (taking $a = -a_3$ or a_1) in case d_1 or d_2 is a p-adic square.

An alternative proof for the case in which d is a p-adic square can be obtained by noting that the success of the foregoing construction is a necessary as well as a sufficient condition for f to be a p-adic zero form. Now with d a p-adic square the two equations to be satisfied by a reduce to

$$(aa_1, -a_1 a_2)_p = (-aa_3, -a_1 a_2)_p = 1.$$

These are inconsistent if the necessary condition fails, and both satisfied by $a = -a_3$ if it holds.

4. Quadratic reciprocity

We introduce the conventional prime ∞ by saying that an integer $a \neq 0$ is an ∞-ADIC SQUARE if and only if $a > 0$, and that f is an ∞-ADIC ZERO FORM if and only if it is indefinite. Now we can extend the definitions and results of the preceding sections by allowing p to take the value ∞; Theorems 20 and 21 remain valid except for the case $n \geqslant 5$ of Theorem 21. Formulae (76), (77) and (81) are valid for $p = \infty$, and we have

$$(a, b)_\infty = \begin{cases} 1 & \text{if} \quad ab \neq 0, \ a \text{ or } b > 0, \\ -1 & \text{if} \quad a < 0, b < 0. \end{cases} \tag{82}$$

The main reason for introducing this use of the symbol ∞ is to simplify the important formula

$$\prod_p (a, b)_p = 1. \tag{83}$$

Here p ranges over ∞, 2, 3, 5, ..., and since by (80) the infinite product has at most finitely many factors $\neq 1$, the left member has a meaning.

By using (81) we see that it suffices to prove (83) in the cases in which each of a, b is either -1 or a prime, and by (76) we may suppose $a \neq b$ except in the case $a = b = -1$, which is easily dealt with. Since $(a, b)_p = (b, a)_p$ and the case $a, b = -1, 2$ is trivial, we may suppose b to be an odd prime, and a to be -1, 2, or another odd prime. Using (78)–(80), these three cases reduce, with a change of notation, to

$$(-1 \,|\, p) = (-1)^{\frac{1}{2}p - \frac{1}{2}}, \quad (2 \,|\, p) = (-1)^{[\frac{1}{4}p + \frac{1}{4}]}, \tag{84}$$
and

$$(p \,|\, q)(q \,|\, p) = \begin{cases} 1 & \text{if} \quad p \text{ or } q \equiv 1 \ (\text{mod } 4), \\ -1 & \text{if} \quad p \equiv q \equiv -1 \ (\text{mod } 4). \end{cases} \tag{85}$$

In (84) and (85), which we must now prove, p, q are any two unequal odd primes. (85) is the law of quadratic reciprocity.

The first step in the proof of (84) and (85) is to show that

$$(a \,|\, p) \equiv a^{\frac{1}{2}p - \frac{1}{2}} \ (\text{mod } p). \tag{86}$$

Now (86) is trivial if $p \,|\, a$; also if $a \equiv x^2 \ (\text{mod } p)$ is a quadratic residue, for then $a^{\frac{1}{2}p - \frac{1}{2}} \equiv x^{p-1} \equiv 1$ by Fermat's theorem. In any case Fermat's theorem gives $a^{p-1} \equiv 1$, $a^{\frac{1}{2}p - \frac{1}{2}} \equiv \pm 1 \ (\text{mod } p)$ unless $p \,|\, a$. If the ambiguous sign were $+$ for any a with $(a \,|\, p) = -1$, then the congruence $x^{\frac{1}{2}p - \frac{1}{2}} \equiv 1 \ (\text{mod } p)$ would have more than $\frac{1}{2}p - \frac{1}{2}$ solutions, which is impossible. We note that (86) gives $(84)_1$ on putting $a = -1$; and for other a we proceed to put it into a more generally useful shape.

For any $a \not\equiv 0 \pmod{p}$, define $\epsilon(a), t(a)$ (depending also on p) by

$$a \equiv \epsilon(a)\, t(a) \pmod{p}, \quad \epsilon(a) = \pm 1, 1 \leqslant t(a) \leqslant \tfrac{1}{2}p - \tfrac{1}{2}.$$

We shall deduce from (86) that

$$(a \mid p) = \prod_{x=1}^{\frac{1}{2}p-\frac{1}{2}} \epsilon(ax) \quad \text{for} \quad p \nmid 2a. \tag{87}$$

It is sufficient to prove that the two sides of (87) are congruent modulo p, since each is ± 1. To do this we note that by the definition of $\epsilon(a), t(a)$ we have $t(a) = t(a')$ if and only if $a \equiv \pm a' \pmod{p}$. Hence with $0 < x_1 < x_2 < \tfrac{1}{2}p$ we cannot have

$$t(ax_1) = t(ax_2).$$

This gives us

$$\prod_x x = \prod_x t(ax) = (\tfrac{1}{2}p - \tfrac{1}{2})! \not\equiv 0 \pmod{p},$$

with x ranging from 1 to $\tfrac{1}{2}p - \tfrac{1}{2}$. Now the right member of (87), multiplied by Πx, $= \Pi t(ax)$, reduces modulo p to Πax, or by (86) to $(a \mid p)\Pi x$, whence (87) follows.

On putting $a = 2$ we easily deduce $(84)_2$ from (87); for $\epsilon(2x) = 1$ for $x = 1, \ldots, [\tfrac{1}{4}p]$, -1 for $x = [\tfrac{1}{4}p] + 1, \ldots, \tfrac{1}{2}p - \tfrac{1}{2}$. For the rest of this section we suppose a odd, and put (87) into the additive shape

$$\tfrac{1}{2} - \tfrac{1}{2}(a \mid p) \equiv \sum_{x=1}^{\frac{1}{2}p-\frac{1}{2}} (\tfrac{1}{2} - \tfrac{1}{2}\epsilon(ax)) \pmod{2}, \quad p \nmid 2a. \tag{88}$$

To simplify (88) we consider the least positive residue of ax modulo p, which is $ax - p[p^{-1}ax]$, and its numerically least residue, which is $\epsilon(ax)\,t(ax)$. These differ by p, 0 in the cases $\epsilon(ax) = -1, 1$; thus we have

$$\tfrac{1}{2}p(1 - \epsilon(ax)) = ax - p[p^{-1}ax] - \epsilon(ax)\,t(ax),$$

giving

$$\tfrac{1}{2} - \tfrac{1}{2}\epsilon(ax) \equiv x + t(ax) + [p^{-1}ax] \pmod{2}.$$

Using again the fact previously noticed that $t(ax)$ ranges with x over the values $1, \ldots, \tfrac{1}{2}p - \tfrac{1}{2}$ (in some order), we see from this formula and (88) that

$$\tfrac{1}{2} - \tfrac{1}{2}(a \mid p) \equiv \sum_{x=1}^{\frac{1}{2}p-\frac{1}{2}} [p^{-1}ax] \pmod{2}.$$

Taking $a = q$, q an odd prime $\neq p$, we see from this last formula that

$$\tfrac{1}{2} - \tfrac{1}{2}(q \,|\, p) \equiv \sum_{x=1}^{\frac{1}{2}p - \frac{1}{2}} \sum_{y=1}^{[p^{-1}qx]} 1 \pmod 2.$$

Here the range of values of y can be taken either as $0 < py \leqslant qx$ or as $0 < py < qx$, since $py = qx$ is impossible; and we have always $y < \tfrac{1}{2}q$ since $x < \tfrac{1}{2}p$. So on interchanging p, q and adding we find

$$\tfrac{1}{2}(p \,|\, q) - \tfrac{1}{2}(q \,|\, p) \equiv \sum_{x} \sum_{y}{}' 1 + \sum_{x} \sum_{y}{}'' 1 = \sum_{x} \sum_{y} 1 \pmod 2,$$

where x, y range from 1 to $\tfrac{1}{2}p - \tfrac{1}{2}$, $\tfrac{1}{2}q - \tfrac{1}{2}$ and $'$, $''$ mean

$$py \leqslant qx, \quad py > qx$$

respectively. The last double sum is, however, $\tfrac{1}{4}(p-1)(q-1)$, which is odd if and only if $p \equiv q \equiv -1 \pmod 4$, whence (85) follows.

5. Necessary and sufficient conditions for representation of zero

We use formula (83) to prove

THEOREM 22. *An integral quadratic form represents zero rationally if and only if it is a p-adic zero form for every p (i.e. for $p = \infty$, 2, 3, 5, ...).*

Proof. The 'only if' is trivial. The 'if' is vacuous for $n = 1$, since a unary form is never a p-adic zero form. By Theorem 20 (i) and a remark preceding Theorem 14, the 'if' for $n = 2$ reduces to showing that d is a rational square, given that d is a p-adic square for every p. This is trivial since the hypothesis gives $d > 0$ and $p^{2u} \| d$ for $p = 2, 3, 5, \ldots$ with integral $u = u(p) \geqslant 0$. The 'if' is also true for $n = 3$, as we have in Theorem 14 proved an apparently weaker condition sufficient.

We next note that the theorem will follow for $n \geqslant 5$ if we can prove it for $n = 4$. We assume as in the proof of Theorem 21 that $f = a_1 x_1^2 + \ldots + a_n x_n^2$, $n \geqslant 5$. Since f is indefinite (i.e. an ∞-adic zero form) the a_i are not all of the same sign and we may suppose $a_1 a_2 < 0$. The ternary form $a_1 x_1^2 + a_2 x_2^2 + a_3 x_3^2$ is therefore an ∞-adic zero form. If there is a p ($\neq \infty$) for which this ternary form is not a p-adic zero form, then by Theorem 20 (iii) $a_1x_1^2 + a_2x_2^2$ does not represent any integer in the p-adic class of $-a_3$; so

$$-a_1a_3x_1^2 - a_2a_3x_2^2$$

does not represent a p-adic square, whence

$$(-a_1a_3, -a_2a_3)_p = -1,$$

giving $p \mid 2a_1a_2a_3$ by (80). It suffices therefore, by the argument of Theorem 21, to find integers y_1, y_2 so that the discriminant of the quaternary form

$$a_1x_1^2 + a_2x_2^2 + a_3x_3^2 + b_4x_4^2, \quad \text{where} \quad b_4 = a_4y_1^2 + a_5y_2^2,$$

is not a p-adic square for any p with $p \mid 2a_1a_2a_3$. We saw in the proof of Theorem 21 that this is possible, with $\mathbf{y} = \mathbf{y}(p)$, say, for any $p \, (\neq \infty)$. Clearly $\mathbf{y} \equiv \mathbf{y}(p) \pmod{p^r}$, with a suitable large r and for each p dividing $2a_1a_2a_3$, gives what is wanted.

We may now take $n = 4$, and assume, again using the fact that we are considering a rationally invariant property, that

$$f = a_1x_1^2 + \ldots + a_4x_4^2,$$

and that f is a p-adic zero form for every p. We have to show that f is a rational zero form, and we may plainly suppose that neither of

$$f_1 = a_1x_1^2 + a_2x_2^2, \quad f_2 = a_3x_3^2 + a_4x_4^2$$

is a rational zero form. By the argument of Theorem 20 (iii) there is for every p an integer $b_p \neq 0$ such that f_1, f_2 represent integers in the p-adic classes of b_p, $-b_p$ respectively. This means that b_pf_1, $-b_pf_2$ represent p-adic squares, and so

$$(b_pa_1, b_pa_2)_p = (-b_pa_3, -b_pa_4)_p = 1.$$

It is clear that the desired conclusion that f is a rational zero form will follow if we can find an integer $a \neq 0$ such that the two forms $f_1 - ax_5^2$, $f_2 + ax_6^2$, are both zero forms. By the case $n = 3$ already proved, this is so if these two ternary forms, or equivalently the two forms $af_1 - x_5^2$, $-af_2 - x_6^2$ derived from them by multiplying by $\pm a$ and putting $a^{-1}x_5$, $a^{-1}x_6$ for x_5, x_6, are p-adic zero forms for every p. This is so if af_1, $-af_2$ represent p-adic squares, or in other words if

$$(aa_1, aa_2)_p = (-aa_3, -aa_4)_p = 1 \quad \text{for all } p.$$

We choose a in the p-adic class of b_p, defined above, for $p = \infty$ and for $p \mid 2a_1a_2a_3a_4$. This is consistent, by Dirichlet's theorem on primes in a progression, with $a = a'p'$, where all the prime factors of $a' \neq 0$ divide $2a_1a_2a_3a_4$, and p' is prime, with $p' \nmid 2a_1a_2a_3a_4$. Now using (80), the conditions which a is required to satisfy fail, if at all, only for $p = \infty$ or $p \mid 2aa_1a_2a_3a_4$, which is the same as $p = \infty$, $p \mid 2a_1a_2a_3a_4$, or $p = p'$. By the choice of a and the properties of the b_p, the conditions on a hold for $p = \infty$, $p \mid 2a_1a_2a_3a_4$, and so fail if at all only for $p = p'$. But if either of them does fail for just one value of p, we have a contradiction with (83). Thus the theorem is proved.

Part (ii) of Theorem 15, stated in Chapter 2 without proof, follows at once from Theorems 21 and 22. Theorem 15 (i) will follow later from Theorem 37 (iv). It will be convenient, however, to prove first the weaker assertion that *a quaternary form is a p-adic zero form (for $p \neq \infty$) if $p \nmid d$*. For diagonal forms this is clear from the proof of Theorem 21. For any f and odd p it follows on noting that the argument of Theorem 3 enables us to diagonalize f without introducing a factor p into the discriminant. For $p = 2$, we may if $2 \nmid d$ suppose $2 \nmid a_{12}$, whence $2 \nmid d_1 = d(f_1), f_1 = f(x_1, x_2, 0, 0)$. Using the transformation (21), we can take f rationally into $f_1 + f_2, d(f_2)$ odd. It suffices to show that this is a 2-adic zero form. Now f_2 will represent an odd integer a, and it is enough to show that $f_1 \equiv -a \pmod 8$ is soluble. This is quite straightforward.

From what we have just proved and Theorem 14, we see that it suffices in Theorem 22 to take $p = \infty$ if $n \geqslant 5$, $p = \infty$ or $p \mid d$ if $n = 4$, $p \mid d$ if $n = 3$; but no such simplification is possible if $n = 2$.

Although, as just mentioned, Theorem 15 (i) will follow later from Theorem 37 (iv), we prove it here as the proof is now very simple. It suffices to show that $d \mid a_{11}$ implies for $p \mid d$ that f (with $n = 4$) is a p-adic zero form. By Theorem 21 we may suppose $p^2 \mid d$, and by Theorem 12, $a_{13} = a_{14} = 0$. From these conditions and $d \mid a_{11}$, we see that if $p \mid a_{12}$ the form $f(p^{-1}x_1, x_2, x_3, x_4)$, with discriminant $p^{-2}d$, satisfies similar conditions. So by introducing induction on $|d|$ we may suppose $p \nmid a_{12}$. We deduce from this that $f_1 = f(x_1, x_2, 0, 0)$ is a p-adic zero form, whence f is such a form. For we have $d(f_1) \equiv a_{12}^2 \pmod{4p}$ a p-adic square.

6. The p-adic rational invariants $\epsilon_p(f)$

It simplifies the definition of these invariants to prove first:

THEOREM 23. *The product*

$$\prod_{1 \leqslant i < j \leqslant n} ((-1)^{i-1}a_i, \, (-1)^j a_j)_p \tag{89}$$

is invariant under all rational transformations under which the form $a_1 x_1^2 + \ldots + a_n x_n^2$ remains diagonal.

Proof. The case $n = 1$ is trivial; the product (89), which we denote for the present by $\epsilon_p(a_1, \ldots, a_n)$, is empty and so has the value 1. The case $n = 2$ is also easy; we have $\epsilon_p(a_1, a_2) = (a_1, a_2)_p$, which is ± 1 according as the form has or has not the rationally invariant property of representing a p-adic square.

The result is trivial for the case in which the transformation interchanges two consecutive a_i, leaving the others unaltered. It is also true for a transformation whose matrix is of the shape $[R, I]$, $R: 2 \times 2$, $I: (n-2) \times (n-2)$, $|R| \neq 0$. For we have

$$\epsilon_p(a_1, \ldots, a_n) = \epsilon_p(a_1, a_2)\, \epsilon_p(a_3, \ldots, a_n)\, (-a_1 a_2, d_2)_p, \tag{90}$$

where $d_2 = (-a_3)\, a_4 \ldots (-1)^n a_n$, which by (10) is

$$(-1)^n d(a_3 x_3^2 + \ldots + a_n x_n^2)$$

apart from a square factor. Now the transformation considered leaves $\epsilon_p(a_1, a_2)$ unaltered as we have just seen, and it replaces $-a_1 a_2 = \tfrac{1}{4} d(a_1 x_1^2 + a_2 x_2^2)$ by $-a_1 a_2 |R|^2$.

The proof will be complete if we can show that, given any diagonal form $b_1 x_1^2 + \ldots + b_n x_n^2$ which is rationally related to $f = a_1 x_1^2 + \ldots + a_n x_n^2$ (with $n \geqslant 3$) we can take f into

$$b_1 x_1^2 + \ldots + b_n x_n^2$$

by a sequence of rational transformations each of which is one of the special kinds just dealt with. We prove this result by induction on n; it is convenient to use the trivial case $n = 2$ as the starting point. We begin by appealing to Theorem 6, and the inductive hypothesis, to show that it is sufficient to take f, by a sequence of transformations of the kinds mentioned, into some form, say $b_1 x_1^2 + c_2 x_2^2 + \ldots + c_n x_n^2$, with leading coefficient b_1. For if so, then $b_2 x_2^2 + \ldots$ and $c_2 x_2^2 + \ldots$ are rationally related.

Now by Theorem 3 we must have
$$b_1 q^2 = a_1 t_1^2 + \ldots + a_n t_n^2,$$
for integers q ($\neq 0$) and $t_1, \ldots, t_n$. The desired result follows at once from the inductive hypothesis—using permutations—if any of the t_i is zero. The general case reduces to this one by the transformation
$$x_1 = t_1 y_1 - a_2 t_2 y_2, \quad x_2 = t_2 y_1 + a_1 t_1 y_2, \quad x_i = y_i \quad (i > 2),$$
which takes f into
$$(a_1 t_1^2 + a_2 t_2^2) y_1^2 + a_1 a_2 (a_1 t_1^2 + a_2 t_2^2) y_2^2 + a_3 y_3^2 + \ldots.$$
Thus the proof is complete apart from the case in which the determinant $a_1 t_1^2 + a_2 t_2^2$ of this transformation vanishes. In this case we lose nothing by supposing $t_1 = t_2 = 0$, and we can appeal to the inductive hypothesis after interchanging a_1, a_n.

We now define the p-ADIC RATIONAL INVARIANT $\epsilon_p(f)$, for every f with $d \neq 0$ and for $p = \infty, 2, 3, \ldots$, to be the value of the product (89) for any diagonal form rationally related to f. It is clear from (90) that
$$\epsilon_\infty(a_1, \ldots, a_n) = \begin{cases} \epsilon_\infty(a_3, \ldots, a_n) & \text{if} \quad a_1 a_2 < 0, \\ -\epsilon_\infty(-a_1, -a_2, a_3, \ldots, a_n) & \text{if} \quad a_1 a_2 > 0, \end{cases}$$
for $n \geqslant 3$. Using this, we easily find
$$\epsilon_\infty(f) = (-1)^{[\frac{1}{4}s + \frac{1}{4}]}; \tag{91}$$
that is, we have $\epsilon_\infty(f) = 1$ if $s \equiv 0, \pm 1$ or $2 \pmod 8$, -1 otherwise. $\epsilon_\infty(f)$ is by (91) a redundant rational invariant; it is introduced to simplify the formula
$$\prod_p \epsilon_p(f) = 1, \tag{92}$$
which follows immediately from (83) (and is to be interpreted in the same way).

We can, for any positive $n_1 < n$, break the product (89) up into three, one with $j \leqslant n_1$, one with $i > n_1$, and a third with $i \leqslant n_1 < j$. Using (10) this gives
$$\epsilon_p(f_1 + f_2) \epsilon_p(f_1) \epsilon_p(f_2)$$
$$= \left. \begin{cases} (d(f_1), d(f_2))_p & \text{if} \quad n(f_1) + n(f_2) \text{ is even,} \\ (d(f_1), -d(f_2))_p & \text{if} \quad n(f_1) \text{ is even and } n(f_2) \text{ odd.} \end{cases} \right\} \tag{93}$$

Using (10), (76), (81) and (89) we find, for $a \neq 0$,

$$\epsilon_p(af) = \begin{cases} \epsilon_p(f) & \text{for odd } n, \\ (a,d)_p \epsilon_p(f) & \text{for even } n. \end{cases} \tag{94}$$

THEOREM 24. *Suppose that either $n = 3$ or $n = 4$ and d is a p-adic square. Then f is a p-adic zero form if and only if $\epsilon_p(f) = 1$.*

Proof. Take first the case $n = 3$; and take f to be a diagonal form $a_1 x_1^2 + a_2 x_2^2 + a_3 x_3^2$. If the product (89) has the value 1 we find, with a little calculation, $(-a_1 a_3, -a_2 a_3)_p = 1$, whence $-a_1 a_3 x_1^2 - a_2 a_3 x_2^2$ represents a p-adic square, and so $a_1 x_1^2 + a_2 x_2^2$ represents an integer in the p-adic class of $-a_3$, and by Theorem 20 (iii) f is a p-adic zero form. The converse is proved similarly.

Now suppose

$$f = \phi + a_4 x_4^2, \quad \phi = a_1 x_1^2 + a_2 x_2^2 + a_3 x_3^2,$$

with $d = 16 a_1 a_2 a_3 a_4 = -4 a_4 d(\phi)$ a p-adic square. Then the product (89) is

$$\epsilon_p(\phi) \, (-a_1 a_2 a_3, a_4)_p = \epsilon_p(\phi) \, (d, a_4)_p = \epsilon_p(\phi).$$

And f is a p-adic zero form, by Theorem 20 (iii), if and only if ϕ represents an integer in the p-adic class of $-a_4$, or of $d(\phi)$; which by Theorem 20 (ii) is the case if and only if ϕ is a p-adic zero form. Thus the assertion for $n = 4$ follows from that for $n = 3$.

We conclude this section by proving that

$$\epsilon_p(f) = 1 \quad \text{if} \quad p \neq \infty \text{ and } p \nmid d. \tag{95}$$

This makes more precise the assertion, to be considered as implied by (92), that the infinite product in (92) has a meaning (cf. (80) and (83)). For odd p and diagonal f (95) follows from (80) and (89). For odd p and general f, it follows by an argument used in §5 in connexion with 4-ary forms. For $p = 2$, we use the decomposition (21), as in the proof just referred to, and formula (93). If $n(f_1) = 2$ and $d(f_1) d(f_2)$ is odd, then $d(f_1) \equiv 1 \pmod 4$ by (52). Hence the symbol $(d(f_1), \pm d(f_2))_2$ occurring in (93) with $p = 2$ is 1 by (79). So the proof of (95) is reduced to the cases $p = 2$, $n = 1, 2$. In case $n = 1$ we have $\epsilon_p(f) = 1$ always as already noted. In case $n = 2$ we have $\epsilon_p(f) = (a_1, -a_1 a_2)_p = (a_1, d)_p$ for diagonal f, whence $\epsilon_p(f) = (a_{11}, d)_p$ unless $a_{11} = 0$. With $p = 2$ this expression is 1 for odd $a_{11}d$, while if a_{11} is even and d odd we have $d \equiv 1 \pmod 8$.

7. Necessary and sufficient conditions for rational relatedness

We are now ready to prove:

THEOREM 25. *Two n-ary forms f, f' with discriminants d, d' (neither zero) are rationally related if and only if dd' is a perfect square, $\epsilon_p(f) = \epsilon_p(f')$ for $p = \infty, 2, 3, \ldots$, and $s(f) = s(f')$.*

Proof. All the conditions are clearly necessary. In case $n = 1$ the condition on dd' is alone sufficient, since it makes

$$f(x_1) = f'(d^{\frac{1}{2}}d'^{-\frac{1}{2}}x_1).$$

We complete the proof of sufficiency by induction on n.

Consider first the case in which f, f' have the same leading coefficient $a \neq 0$. We can split off ax_1^2 rationally from each of f, f', and so suppose without loss of generality that $f, f' = ax_1^2 + g$, $ax_1^2 + g'$. It suffices to show that g, g' satisfy the conditions of the theorem and so are rationally related. This is quite straightforward (use (9) and (93)), and the obvious

$$s(ax_1^2 + g) = \operatorname{sgn} a + s(g)).$$

To complete the proof it suffices to show that the conditions imply that there is an integer $a \neq 0$ represented by each of f, f'. This is trivial if either of f, f' is a zero form (the zero form (53) represents all multiples of a_{12}^2). In other cases it easily follows if we can show that the $2n$-ary form $f - f'$ is a zero form. As this form has signature 0, it suffices by Theorem 15 to consider the case $n = 2$. In this case we can verify easily by using (93) that $f - f'$ satisfies the conditions of Theorem 22.

We may regard Theorem 25 as asserting that the system of rational invariants so far found is complete; that is, that any other rational invariant must be a function of these. In the opposite direction, we can show that the system is irredundant except for the relations already noticed. It simplifies the argument to ignore the signature at first and prove:

THEOREM 26. *In order that there may exist a form f of given rank n, whose invariants $\epsilon_p(f)$ have given values ϵ_p (each ± 1), and such that $d_0 d(f)$ is a perfect square, $\neq 0$, for some given $d_0 \neq 0$, it is necessary and sufficient that*

(i) $\prod\limits_{p} \epsilon_p = 1$ *(that is, the number of negative ϵ_p is finite and even);*

(ii) $\epsilon_p = 1$ *for every p if $n = 1$, and for every p for which d_0 is a p-adic square if $n = 2$.*

Proof. Necessity and sufficiency are both trivial for $n = 1$. The necessity of (i) is clear from (92) in all cases.

For $n = 2$ we choose a suitable $a \neq 0$ and take $f = ax_1^2 - ad_0 x_2^2$. This gives $d(f) = 4a^2 d_0$, and, by (76),

$$\epsilon_p(f) = (a, -ad_0)_p = (a, d_0)_p.$$

The necessity of (ii) is clear from this; for there is no loss of generality in taking $f = ax_1^2 + bx_2^2$, and then $-abd_0$ has to be a square.

Assuming (i) and (ii), sufficiency follows if we can find a such that

$$(a, d_0)_p = \epsilon_p \quad \text{for all } p. \tag{96}$$

The proof that this is possible is similar to, but simpler than, that of Theorem 22. Condition (ii) ensures that (96) can be satisfied for any given p. (i) and (83) ensure that it fails, if at all, for evenly many and so for at least two values of p. On the other hand, we can choose a so that (96) fails for at most one value of p.

For $n \geqslant 3$ we proceed by induction. We seek a form f of the shape $ax_1^2 + g$. If $n \geqslant 4$ we find that a suitable g can be constructed for any $a \neq 0$. Hence the details of the proof are given only for the case $n = 3$. Supposing a suitably chosen, the conditions on g to give an f with the required properties are like those on f, but with ad_0 for d_0 and

$$\eta_p = \epsilon_p(-a, ad_0)_p \quad \text{for} \quad \epsilon_p.$$

It is clear that the conditions (i) and (ii) are unaltered by putting ad_0 for d_0, $2 = n - 1$ for $n = 3$, and η_p for ϵ_p provided only that $\eta_p = 1$ for every p for which ad_0 is a p-adic square. For such p, however, we have $\eta_p = \epsilon_p$. All we have to do therefore is to choose a in a p-adic class other than that of d_0 for every p for which $\epsilon_p = -1$.

THEOREM 27. *The form f whose existence is asserted in Theorem 26 can be chosen to have signature s if and only if $|s| \leqslant n$, $s \equiv n \pmod 2$, $(-1)^{[\frac{1}{2}s]} = \operatorname{sgn} d_0$, and $(-1)^{[\frac{1}{4}s+\frac{1}{4}]} = \epsilon_\infty$.*

Proof. The necessity of these conditions on s is clear from (23) and (91). The last three of them determine the residue of s modulo 8, and then the first of them determines s uniquely if $n \leqslant 3$. There is therefore nothing more to prove unless $n \geqslant 4$. Now a slight modification of the argument of Theorem 26 gives the result (restrict a to have the sign of s if $|s| = n$).

CHAPTER 4

p-ADIC EQUIVALENCE

1. Definition; invariance of $\epsilon_p(f)$

We shall say that f is p-ADICALLY EQUIVALENT to f', or in symbols $f \underset{p}{\sim} f'$, if for every positive integer t there exists a matrix P with

$$f^P \equiv f' \pmod{p^t}, \quad p \nmid |P|, \quad P \text{ integral}. \tag{97}$$

If P satisfies conditions $(97)_2$ and $(97)_3$ there is a Q satisfying the same conditions and with $PQ \equiv I \pmod{p^t}$; we can take $Q = hqP^{-1}$, with h chosen to make $hq \equiv 1 \pmod{p^t}$, for any q (say $|P|$) such that $p \nmid q$ and qP^{-1} is integral. From this remark it is clear that the relation of p-adic equivalence has the properties (16)–(18) discussed in Chapter 1. The relation is important because of the obvious fact that (for every p) $f \underset{p}{\sim} f'$ is a necessary condition for $f \sim f'$.

Three obvious invariants under this relation are (i) the rank n, (ii) the integer, say u, such that $p^u \| d$, and (iii) the p-adic class of d. (Take $t = u + 3$ and note that (97) makes $d(f)|P|^2 \equiv d(f')$ $\pmod{p^t}$.) Another is given by

THEOREM 28. $\epsilon_p(f)$ *is invariant under p-adic equivalence.*
Proof. Write for brevity, for $i = 1, ..., n$,

$$f_i = f(x_1, ..., x_i, 0, ..., 0), \quad d_i = d(f_i); f_n = f,$$
$$d_n = d\ (\neq 0), \quad d_1 = a_{11}.$$

An easy induction on n—using (38)—shows that (i) every form is equivalent to one with $d_1...d_{n-1} \neq 0$ and (ii) if $d_1...d_{n-1} \neq 0$ then f is rationally related to the form

$$d_1 x_1^2 - d_1 d_2 x_2^2 + d_2 d_3 x_3^2 - ... - + (-1)^{n-1} d_{n-1} dx_n^2,$$

with discriminant $(d_1...d_{n-1})^2 d$. Now from (89), using (77) and (81), a simple calculation gives

$$\epsilon_p(f) = \prod_{1 \leqslant i < n} (d_i, d_{i+1})_p. \tag{98}$$

Now we know already from its definition that $\epsilon_p(f)$ is rationally invariant, whence $\epsilon_p(f^P) = \epsilon_p(f)$. Hence we have only to show that addition of any multiples of p^t to the coefficients will not alter $\epsilon_p(f)$ if t is chosen sufficiently large. And it is sufficient to prove this for a form equivalent to f and with $d_1 \ldots d_{n-1} \neq 0$. So we may use (98); and the desired result is clear if we choose t so that p^{t-2} divides none of the d_i (see (77)). This completes the proof.

The main objects of this chapter are to show (i) that in the case $p \nmid d$ all the invariants under p-adic equivalence are determined by n and d and (ii) that in any case it suffices to give t in (97) a value depending only on $d(f)$. We also consider the related problem of determining whether, for a given integer a, the congruence $f(\mathbf{x}) \equiv a \pmod{p^t}$ is soluble, with or without the restriction $p \nmid \mathbf{x}$. Here again we shall see that it suffices to consider a finite number of prime powers p^t.

2. The case $p \nmid d$

It is convenient to consider the cases of odd, even n separately.

THEOREM 29. *Suppose that n is odd and $p \nmid d$. Then*

$$f \underset{p}{\sim} x_1 x_2 + \ldots + x_{n-2} x_{n-1} + d x_n^2. \tag{99}$$

Suppose further that $p \neq 2$, and let $\mathbf{x}$ take p^n values, no two congruent modulo p; then $(f(\mathbf{x}) \mid p)$ takes the value $(d \mid p)$ more often than $-(d \mid p)$.

Proof. The first assertion is clear in case $n = 1$ (the right member of (99) is then to be taken to mean $d x_1^2$). We proceed by induction from $n - 2$ to n, assuming $n \geqslant 3$, which gives $\epsilon_p(f) = 1$ by (95), whence f is a p-adic zero form by Theorem 24. Now as in the proof of Theorem 20 (ii) we may suppose

$$p^t \mid a_{11}, \quad a_{13} = \ldots = a_{1n} = 0.$$

With $p \nmid d$ it follows that $p \nmid a_{12}$. We split off the binary form $a_{11} x_1^2 + a_{12} x_1 x_2 + a_{22} x_2^2$ by a transformation whose matrix is integral and, with $a_{12}^2 - 4 a_{11} a_{22}$, prime to p.

The binary form we have split off can be taken into one congruent to $x_1 x_2$ modulo p^t by putting $h x_1$ for x_1, $h a_{12} \equiv 1 \pmod{p^t}$,

4-2

and then making an obvious parallel transformation. Thus we may suppose (for any t) $f \equiv x_1 x_2 + g(x_3, \ldots, x_n) \pmod{p^t}$; and this process clearly leads us to (99).

The second assertion is clearly invariant under p-adic equivalence, so it need only be proved for the case in which f is equal to the right member of (99). That is, it suffices to prove

$$\sum_{x_1=0}^{p-1} \cdots \sum_{x_n=0}^{p-1} (x_1 x_2 + \ldots + d x_n^2 \,|\, p)(d\,|\,p) > 0.$$

Now if we give $x_2, \ldots, x_n$ any fixed values with $p \nmid x_2$, then summation over x_1 gives zero, as f takes $\frac{1}{2}p - \frac{1}{2}$ quadratic residue and $\frac{1}{2}p - \frac{1}{2}$ non-residue values. Hence the multiple sum is equal to p times the corresponding sum over $x_3, \ldots, x_n$, with the term $x_1 x_2$ omitted; repeating this argument, its value is $p^{\frac{1}{2}n - \frac{1}{2}}(p - 1)$.

THEOREM 30. (i) *Suppose that n is even, $p \nmid d$, and d is a p-adic square. Then*

$$f \underset{p}{\sim} x_1 x_2 + \ldots + x_{n-1} x_n, \tag{100}$$

and $f(\mathbf{x}) \equiv 0 \pmod{p}$ has more than p^{n-1} solutions.

(ii) *Suppose that n is even, $p \nmid d$, and d is not a p-adic square. Then*

$$f \underset{p}{\sim} x_1 x_2 + \ldots + x_{n-3} x_{n-2} + x_{n-1}^2 + d x_{n-1} x_n + \tfrac{1}{4} d(1 - d) x_n^2, \tag{101}$$

and $f(\mathbf{x}) \equiv 0 \pmod{p}$ has fewer than p^{n-1} solutions.

Proof. (i) is proved in the same way as Theorem 29. Instead of summing the Legendre symbol we use a summand which takes the values -1 if $p \nmid f$, $p - 1$ if $p \mid f$.

The arguments used for (i) and Theorem 29 reduce the proof of (ii) to the case $n = 2$. In this case the second assertion is easier to prove. If $f(x_1, x_2) \equiv 0 \pmod{p}$ has any solution other than the trivial $x_1 \equiv x_2 \equiv 0 \pmod{p}$ then we may suppose $p \mid a_{11}$ and this makes $d \equiv a_{12}^2 \pmod{4p}$ a p-adic square. The number of solutions of $p \mid f$ is thus $1 < p = p^{n-1}$.

It remains therefore only to prove (101) for the case $n = 2$. We shall not use the hypothesis that d is not a p-adic square. Thus we shall see that (101) holds also in case (i), though it is then less convenient than (100). Now the desired result is clear for forms with leading coefficient 1. For with $a_{11} = 1$ (and $a_{12} \equiv d \pmod 2$)

we can replace the coefficient a_{12} by d by putting $x_1 + \frac{1}{2}(d - a_{12})x_2$ for x_1. It is therefore sufficient to show that $f(\mathbf{x}) \equiv 1 \pmod{p^t}$ is soluble for every t.

This reduces to solving $(f(\mathbf{y})|p) = 1$, or $f(\mathbf{y}) \equiv 1 \pmod 8$ if $p = 2$. For then we can put $\mathbf{x} = h\mathbf{y}$ and solve $h^2 f(\mathbf{y}) \equiv 1 \pmod{p^t}$. It is easily verified that with $2 \nmid d$ we can solve $f(\mathbf{y}) \equiv 1 \pmod 8$. With $p \neq 2$, we may, by using the transformation (20), take f into $a_{11}x_1^2 - a_{11}dx_2^2$ (supposing as we may that $p \nmid a_{11}$). The argument used to prove (80) shows that this form takes quadratic residue values, and the proof is complete.

It will be convenient to denote by σ a sum of products $x_1 x_2 + x_3 x_4 + \ldots$, which may possibly be empty or have just one term. The number of terms will be clear, by the convention of Chapter 1, §1, regarding disjoint forms, from the context. In (99) we may replace d by one of ± 1, ± 3 if $p = 2$, or by one of 1, b if $p \neq 2$, $b = b_p$ being any fixed quadratic non-residue modulo p. This may be done by putting hx_n for x_n, with suitable h prime to p, and reducing modulo p^t. Similarly in (101) we may replace d by -3 if $p = 2$, or $4b$ if $p \neq 2$. Thus the right member of (101) may be replaced by $\sigma + \psi$, where $\psi = \psi_p$ is defined by

$$\left.\begin{aligned}
\psi_2 &= x_1^2 + x_1 x_2 + x_2^2, \\
\psi_p &= x_1^2 - bx_2^2 \quad (p \neq 2),
\end{aligned}\right\} \tag{102}$$

with b as above. (To do this we must in (101) replace x_{n-1} by $x_{n-1} + hx_n$, $2h \equiv -d \pmod{p^t}$ if $p \neq 2$, $h = 2$ if $p = 2$.) Thus the forms (99)–(101) reduce to

$$\sigma + cx_n^2, \quad \sigma, \quad \sigma + \psi_p, \tag{103}$$

with ψ_p as in (102) and $c = 1$ or b if p is odd, ± 1 or ± 3 if $p = 2$.

From Theorem 30 (ii) we see that

$$h\psi_p \underset{p}{\sim} \psi_p \quad \text{if} \quad p \nmid h. \tag{104}$$

From this it easily follows that $\psi_p \equiv h \pmod p$ has just as many solutions as $\psi_p \equiv 1 \pmod p$ for $p \nmid h$; whence the hypothesis that n is odd is necessary in the second part of Theorem 29. Similarly, the assertions in Theorem 30 regarding the congruence $f(\mathbf{x}) \equiv 0 \pmod p$ are false for odd n.

In general, a non-p-adic zero form f is one such that $p^t|f(\mathbf{x})$ implies $p\,|\,\mathbf{x}$ for some t. (102) shows that $\psi = \psi_p$ has this property for $t = 1$; that is, $p\,|\,\psi(\mathbf{x})$ implies $p\,|\,\mathbf{x}$, whence it implies $p^2|\psi(\mathbf{x})$. On the other hand, (104) shows that ψ behaves well in congruences modulo a power of p with residue not divisible by p; $\psi(\mathbf{x}) \equiv a \pmod{p^t}$ is always soluble if $p \nmid a$.

3. The case $p \neq 2$

The investigation of 2-adic equivalence becomes rather complicated when $2\,|\,d$. We therefore deal with odd p first, and begin with a classical result:

THEOREM 31. *For odd p, every form is p-adically equivalent to a diagonal form.*

Proof. It is enough to show that a unary form can if $n \geqslant 2$ be split off rationally from f by a transformation with integral coefficients and determinant prime to p. It is clearly sufficient to prove this for primitive f. So by an equivalence transformation we may suppose $p \nmid a_{11}$; then Theorem 2 gives what is wanted. We can of course take each diagonal coefficient to be either a prime power p^r or bp^r, with $b = b_p$ as in §2.

It is desirable to coarsen this decomposition, introducing into it as many product terms as possible, since these have much simpler properties than squares.

THEOREM 32. *Every form f has, for each odd p, a unique p-adic decomposition*

$$f \underset{p}{\sim} f^{(0)} + pf^{(1)} + \ldots + p^r f^{(r)} + \ldots, \tag{105}$$

where each $f^{(r)}$ is either identically 0 or is one of the standard forms (103).

Remarks. There can of course be only finitely many terms not identically 0 in (105); and there are some obvious special cases in which (105) has only one summand and so is not strictly a decomposition. There should logically be a reference to p in the notation for the $f^{(r)}$, which depend on p; but for simplicity it is omitted.

Proof of Theorem 32. We use Theorem 31, and gather together, for each $r \geqslant 0$, all diagonal terms whose coefficients are divisible exactly by p^r. Now on appealing to Theorems 29, 30, we obtain the desired decomposition and have only to prove its uniqueness. What we have to prove is (i) that n_r, the rank of $f^{(r)}$, is a p-adic invariant of f for every $r \geqslant 0$ and (ii) that $(d_r|p)$ $(= \pm 1)$ is such an invariant for every r with $f^{(r)}$ not identically 0, where $d_r = d(f^{(r)})$.

The proofs depend on the fact that the number of solutions of the vector congruence $A\mathbf{x} \equiv \mathbf{0}$ (mod p^r) is an invariant of the form f, with matrix A, under p-adic equivalence. To prove this, suppose $\mathbf{x} \equiv P\mathbf{y}$ (mod p^r), $p \nmid |P|$, and let $f' \equiv f^P$ (mod p^r) have matrix $B \equiv P'AP$ (mod p^r). Then clearly $B\mathbf{y} \equiv P'AP\mathbf{y} \equiv \mathbf{0}$ (mod p^r) if and only if $AP\mathbf{y} \equiv A\mathbf{x} \equiv \mathbf{0}$ (mod p^r).

If f is equal to the right member of (105), denote by A_r the matrix of $f^{(r)}$. Then $|A_r| = \pm d(f^{(r)})$ or $\pm 2d(f^{(r)}) \not\equiv 0$ (mod p). The congruence $A\mathbf{x} \equiv 0$ (mod p^r) thus reduces to

$$p^{r-u}|\mathbf{x}_u \quad \text{for} \quad 0 \leqslant u \leqslant r-1, \tag{106}$$

where $\mathbf{x}_u$ is the vector argument of $f^{(u)}$. The number of solutions of $A\mathbf{x} \equiv \mathbf{0}$ (mod p^r) is thus a power of p, with exponent

$$rn - \sum_{0 \leqslant u < r} (r-u)\, n_u.$$

By the invariant property just proved, this holds generally (i.e. without assuming equality in (105)). It follows that the expression just written is an invariant of f under p-adic equivalence. Subtracting the expression derived by putting $r-1$ for r, we see that (for $r \geqslant 1$) $n - (n_0 + \ldots + n_{r-1})$ is invariant; whence differencing again it is clear that n_r is so, as was to be proved.

To deal with the $(d_r|p)$, we consider the values (all integral by (3)) of $p^{-r}f(\mathbf{x})$ for $\mathbf{x}$, with each element ranging from 0 to $p^{r+1}-1$, satisfying $A\mathbf{x} \equiv \mathbf{0}$ (mod p^r). First suppose that n_r is odd. We shall sum the Legendre symbol $(p^{-r}f(\mathbf{x})|p)$ over these $\mathbf{x}$. The sign of the sum is clearly invariant under p-adic equivalence; and we shall prove that it is $(d_r|p)$. This assertion follows from Theorem 29 on noting that (106) gives $p^{r+1}|\dot{p}^{2r-u}|p^u f^{(u)}$ for $u < r$, whence it implies

$$f^{(0)} + pf^{(1)} + \ldots p^r f^{(r)} + \ldots \equiv p^r f^{(r)} \text{ (mod } p^{r+1}). \tag{107}$$

Similarly, if n_r is even and positive, we see from Theorem 30 and (107) that the proportion of the $\mathbf{x}$ with $0 \leqslant x_i < p^{r+1}$ and

$p^r \mid A\mathbf{x}$ for which $p^{r+1} \mid f(\mathbf{x})$ is never equal to p^{-1}. The sign of the difference, which is clearly invariant, is that of $(d_r \mid p)$; and this completes the proof.

4. Conditions for p-adic equivalence

If $f \underset{p}{\sim} f'$ it is clear that

$$n(f) = n(f'), \quad d(f)\,d(f') \text{ is a } p\text{-adic square}, \tag{108}$$

and that

$$p^u \parallel d(f), \quad p^u \parallel d(f'), \tag{109}$$

for some $u = u(p,f) \geqslant 0$. We prove:

THEOREM 33. (i) *If the forms f, f' satisfy (97) with some P and t such that*

$$p^{t-1} \parallel 4d(f), \tag{110}$$

then $f \underset{p}{\sim} f'$.

(ii) *If f, f' satisfy conditions (108) and (109) then we have $f \underset{p}{\sim} f'$ if also (97) can be satisfied with u in place of t.*

Proof. It is not difficult to see that (97), with the value of t given by (110), implies (108) and (109); hence (i) follows from (ii). From the definition in §1 it is clear that (ii) follows if we can show that (108), (109) and

$$f \equiv f' \pmod{p^u} \tag{111}$$

imply $f \underset{p}{\sim} f'$.

First suppose that $p \neq 2$. In determining $f^{(r)}$, in the proof of Theorem 32, we worked with congruences modulo p^v, with v always at most $r+1$. Hence with an obvious temporary notation based on that of (105), we see that $f^{(r)}(f) = f^{(r)}(f')$ for $r = 0, \ldots, u-1$. Clearly this also holds (both sides identically 0) for $r > u$; hence also, using (108) and (109), for $r = u$. This gives the result.

Now let $p = 2$. We use a different method, because the analogue of Theorem 32 for $p = 2$ is rather complicated. The case $n = 1$ is trivial, so we may assume $n \geqslant 2$ and use induction. We may also assume that f is not identically 0 modulo 2. For if $f = 2f_1$, f_1 integral, then (111) implies $f' = 2f_1'$, f_1' integral, and $f_1 \equiv f_1'$ $\pmod{2^{u-1}}$, whereas $u(2,f_1) = u(2,f) - n < u(2,f) - 1$. We distinguish two cases, (i) $2 \nmid A$, (ii) $2 \mid A$.

In the first case some a_{ij} with $i \neq j$ is odd, so we may suppose

a_{12} odd. We may suppose $u \geqslant 1$, that is, $2 \mid d$. For Theorems 29, 30 give the result if d is odd. With $2 \nmid a_{12}, 2 \mid d$ we have $n \geqslant 3$. We can (see (21)) find P, integral, with $|P|$ odd, such that $f^P = f_1 + f_2$, with $f_1 = f(x_1, x_2, 0, ..., 0)$. Applying the same process to f' and using (111), we see that there exists $Q \equiv P \pmod{2^u}$ such that $f'^Q = f_1' + f_2'$, whence we may assume in place of (111)

$$f = f_1 + f_2 \equiv f_1' + f_2' = f' \pmod{2^u},$$

f_1, f_1' binary, each with odd discriminant. This gives that $(108)_1$, (109) and (111) all hold with f_2, f_2' for f, f', since using (9) we have $u(f_2) = u(f)$.

Now we must show that $f_1 \underset{2}{\sim} f_1'$. This will give us $(108)_2$ with f_1, f_1' for f, f', whence by (9) we see that $(108)_2$ holds also with f_2, f_2' for f, f', and the inductive hypothesis then gives $f_2 \underset{2}{\sim} f_2'$. But the proof of $f_1 \underset{2}{\sim} f_1'$ reduces by Theorem 30 to showing that $d(f_1) d(f_1') \equiv 1 \pmod 8$. That is, we have only to show (using a_{ij}' to denote the general coefficient of f') that

$$(a_{12}^2 - 4a_{11}a_{22})(a_{12}'^2 - 4a_{11}'a_{22}') \equiv 1 \pmod 8.$$

This follows easily from (111) and $2 \nmid a_{12}$.

We now come to the case $2 \mid A$. As we assume $2 \nmid f$, we may suppose a_{11} odd, and $l = \frac{1}{2}a_{12}x_2 + ... + \frac{1}{2}a_{1n}x_n$ is a linear form with integral coefficients. We use the transformation

$$x_1 = y_1 - l, \quad x_i = a_{11}y_i \quad (i = 2, ..., n), \tag{112}$$

to take f into a disjoint form

$$a_{11}x_1^2 + a_{11}^2 f(0, x_2, ..., x_n) - a_{11}l^2.$$

We note that with $n \geqslant 2$ and $2 \mid A$ we have $4 \mid d, u \geqslant 2$, by (7). If we treat f' in the same way as f, then a simple calculation shows that, with an obvious notation, we have $l \equiv l' \pmod{2^{u-1}}, l \equiv -l' \pmod 2$, $l^2 \equiv l'^2 \pmod{2^u}$. That is, we may make f, f' disjoint without disturbing (111), which we may therefore replace by

$$f = a_{11}x_1^2 + f_2 \equiv a_{11}'x_1^2 + f_2' = f' \pmod{2^u}.$$

If $u \geqslant 3$ we have a_{11} odd and congruent to a_{11}' modulo 8, $u(f_2) \leqslant u(f)$ by (9), and the rest of the argument is like that for the case $2 \nmid A$.

We may therefore suppose $u = 2$, that is, $d \equiv 4 \pmod 8$, and this by (7) and $2 \mid A$ gives us $n \leqslant 3$. The cases that remain are simple, though they require special treatment. The next theorem shows that if $n = 2$, $2 \nmid f$, $d \equiv 4 \pmod 8$, and f' satisfies the same conditions, also $(108)_2$, then $f \underset{\sim}{_2} f'$ can be deduced by using instead of (111) the weaker assumption that $f \equiv 1$, $f' \equiv 1 \pmod 4$ are both soluble or both insoluble. Under the same conditions except that $n = 3$, Theorem 34 shows that $f \underset{\sim}{_2} f'$ if neither or both of $4 \mid f(\mathbf{x})$, $4 \mid f'(\mathbf{x})$ imply $2 \mid \mathbf{x}$.

5. 2-adic decomposition

To obtain the analogue for $p = 2$ of Theorem 32, also to complete the proof of Theorem 33, we need besides Theorems 29, 30 also:

THEOREM 34. *Suppose $d \equiv 4 \pmod 8$, $2 \mid A$, and f not identically 0 modulo 2. Then*

(i) *if $n = 2$ and $d \equiv 4 \pmod{16}$ we have $f \underset{\sim}{_2} x_1^2 - \tfrac{1}{4}dx_2^2$;*

(ii) *if $n = 2$ and $d = -4 \pmod{16}$ we have $f \underset{\sim}{_2} \pm(x_1^2 - \tfrac{1}{4}dx_2^2)$ according as $f \equiv 1 \pmod 4$ is or is not soluble;*

(iii) *if $n = 3$ and f is a 2-adic zero form then $f \underset{\sim}{_2} \tfrac{1}{4}dx_1^2 + 2x_2x_3$;*

(iv) *if $n = 3$ and f is not a 2-adic zero form then*

$$f \underset{\sim}{_2} -\tfrac{3}{4}dx_1^2 + 2\psi = -\tfrac{3}{4}dx_1^2 + 2(x_2^2 + x_2x_3 + x_3^2),$$

and $4 \mid f(\mathbf{x})$ implies $2 \mid \mathbf{x}$.

Proof. If $n = 2$ we can suppose f, by the transformation (112), to be a diagonal form $ax_1^2 + bx_2^2$, with $d = -4ab$. Obviously we can take a to be one of ± 1, ± 3. But $d = -4ab \equiv 4 \pmod 8$ makes b odd, so f represents $a + 4b \equiv \pm 1 \pmod 8$ in case $a \equiv \mp 3 \pmod 8$. It follows that we can take $a = \pm 1$ always; and clearly if $ab \equiv -1 \pmod 4$ we can take $a = 1$. (i) and (ii) follow.

Now if $n = 3$, consider first the case in which $f \equiv \tfrac{1}{4}d \pmod 8$ is soluble. In this case we have, using (112),

$$f \underset{\sim}{_2} \tfrac{1}{4}dx_1^2 + f_2, \quad d(f_2) \equiv 4 \pmod{32}.$$

If $2 \mid f_2$ we apply Theorem 30 to the form $\tfrac{1}{2}f_2$, and have

$$f \underset{\sim}{_2} \tfrac{1}{4}dx_1^2 + 2x_2x_3.$$

If not, we apply (i) above to the form $\frac{1}{4}df_2$, and we see that $f_2 \underset{2}{\sim} \frac{1}{4}df_2 \underset{2}{\sim} x_2^2 - x_3^2$, which gives $f \underset{2}{\sim} \frac{1}{4}d(x_1^2 + x_2^2 - x_3^2)$. This (see (29)) gives

$$f \underset{2}{\sim} \frac{1}{4}dx_1^2 + \frac{1}{4}d \cdot 2x_2x_3 \underset{2}{\sim} \frac{1}{4}dx_1^2 + 2x_2x_3.$$

Next suppose $f \equiv -\frac{3}{4}d$ (mod 8) soluble, $f \equiv \frac{1}{4}d$ (mod 8) insoluble. (112) now gives $f \underset{2}{\sim} -\frac{3}{4}dx_1^2 + f_2$, with $f_2 \equiv 4$ (mod 8), whence also $f_2 \equiv 1$ (mod 2), insoluble. Applying Theorem 30 to the form $\frac{1}{2}f_2$, we find $f \underset{2}{\sim} \frac{3}{4}dx_1^2 + 2\psi$. In this case, $4|f$ implies $2|x_1, 2|\psi, 2|x_2, x_3$.

The proof is complete if we obtain a contradiction from the assumption that $f \equiv \frac{1}{4}d, -\frac{3}{4}d$ (mod 8) are both insoluble. In this case (112) shows that $f \underset{2}{\sim} ax_1^2 + f_2$, with $a \equiv -\frac{1}{4}d$ (mod 4), whence $d(f_2) \equiv -4$ (mod 16) by (9). This gives $d(\frac{1}{2}f_2) \equiv -1$ (mod 4), so by (52) $\frac{1}{2}f_2$ is not integral, and by (ii) above we can take f_2, hence f, into a diagonal form, say $ax_1^2 + bx_2^2 + cx_3^2$, with $abc = -\frac{1}{4}d$ odd. We may suppose $a \equiv b \equiv c \equiv -\frac{1}{4}d$ (mod 4); but now we have the desired contradiction on noting that f represents $a+b+c \equiv \frac{1}{4}d$ (mod 4).

The standard forms (103), with $p = 2 \nmid d$, may be written as

$$\sigma \pm cx_n^2, \quad c = 1, -3, n \text{ odd}, \tag{113}$$

(σ denoting as before a sum of products) and

$$\sigma, \quad \sigma + \psi = \sigma + x_{n-1}^2 + x_{n-1}x_n + x_n^2 \quad (n \text{ even}). \tag{114}$$

It is convenient, in the 2-adic analogue of Theorem 32, to use, besides these, six other standard forms, with $d \equiv 4$ (mod 8). These are (with $c = 1, -3$)

$$\sigma + x_{n-1}^2 - cx_n^2, \quad \sigma + x_{n-1}^2 + cx_n^2, \quad \sigma - x_{n-1}^2 - cx_n^2. \tag{115}$$

It is sometimes convenient to replace these (putting $x_{n-1} + x_n$ for x_{n-1}) by

$$\sigma + x_{n-1}^2 + 2x_{n-1}x_n + (1-c)x_n^2, \quad \sigma \pm (x_{n-1}^2 + 2x_{n-1}x_n + (1+c)x_n^2). \tag{116}$$

THEOREM 35. *Every form f has a 2-adic decomposition*

$$f \underset{2}{\sim} f^{(0)} + 2f^{(1)} + \ldots + 2^r f^{(r)} + \ldots, \tag{117}$$

in which every $f^{(r)}$ is a form, of some rank $n_r \leqslant n$, $\geqslant 0$, of one of the shapes (113)–(115). *The n_r and the distinction (for even n_r) between cases* (114) *and* (115) *are unique.*

Proof. The first assertion will follow if we assume that (for any integral $f^{(r)}$) (117) is impossible except with $f^{(r)}$ identically 0 for $r \neq 0$, and deduce that f itself is 2-adically equivalent to one of the forms (113)–(115).

This assumption, which makes f not identically 0 modulo 2, enables us to use the transformations (21) and (112) used in the proof of Theorem 33. Thus we may suppose

$$f = g + \phi,$$

where g is a disjoint sum of binary forms, each with odd discriminant, and ϕ is diagonal (one of g, ϕ may be identically 0). (ϕ has odd coefficients.)

We obtain the desired result by Theorem 29 or 30 unless $n(\phi) \geqslant 2$. On the other hand, $n(\phi) \geqslant 3$ contradicts the assumption made above (see parts (iii) and (iv) of Theorem 34). So we suppose $n(\phi) = 2$,

$$f = g + ax_{n-1}^2 + bx_n^2, \quad ab \text{ odd},$$

with g as above. Applying Theorem 29 to the form $g + ax_{n-1}^2$, we may suppose $g = \sigma$. Now applying part (i) or (ii) of Theorem 34 to $ax_{n-1}^2 + bx_n^2$ the proof of the first assertion is completed.

The proof of uniqueness is like that of Theorem 32. It is convenient to write

$$f^{(r)} = g^{(r)} + \phi^{(r)}, \tag{118}$$

with $n(g^{(r)})$, $= n_r'$, say, even and $d(g^{(r)})$ odd, ϕ being diagonal, with odd coefficients, and rank $n_r'' = 0, 1$ or 2. The matrix, say B_r, of $g^{(r)}$ has $|B_r|$ odd. The matrix of $\phi^{(r)}$ is divisible by 2, and may be written as $2C_r$, $|C_r|$ odd. With this notation the matrix of the right member of (117) is

$$[B_0, 2C_0, \dots, 2^r C_{r-1}, 2^r B_r, \dots].$$

By considering the congruences $A\mathbf{x} \equiv 0 \pmod{2^r}$ we see just as in Theorem 32 that the numbers $n_{r-1}'' + n_r'$ are invariant under 2-adic equivalence.

Since n_r'' takes only the values 0, 1, 2, while n_r' is always even, the proof will be complete if we can show that sgn n_r'' ($= 0$, 1 in cases (114) and (115)) is invariant. To prove this we note that, by an argument like that of Theorem 32, the congruence $2^r | A\mathbf{x}$ implies that 2^{r-u} divides all the variables of $g^{(u)}$, while 2^{r-u-1} divides all those of $\phi^{(u)}$. This means that $2^r | A\mathbf{x}$ implies

$$f(\mathbf{x}) \equiv 2^{r-1}\phi^{(r-1)} \pmod{2^r};$$

whence $2^r | f(\mathbf{x})$ follows if $\phi^{(r-1)}$ is identically 0, otherwise not. Thus we see that $f^{(r)}$ is of the shape (114) if and only if $2^{r+1} | A\mathbf{x}$ implies $2^{r+1} | f(\mathbf{x})$; and this completes the proof.

It is possible to standardize the decomposition (117) so that it becomes unique; but it is not logically necessary to do so, since by Theorem 33 we can always find, by a finite though tedious process of trial and error, whether two given forms f, f' are or are not 2-adically equivalent. The standardization is somewhat tedious, and depends on certain conventions which are to some extent arbitrary. The proof that it makes (117) unique is also tedious.

6. A coarser p-adic decomposition

We shall later need decompositions valid under p-adic equivalence for more than one value of p simultaneously. These will be obtained from:

THEOREM 36. *Suppose that $n \geqslant 8$. Then f has a decomposition*

$$f \underset{p}{\sim} f_1 + f_2, \tag{119}$$

with $n(f_1) = n_1 = 2$ or 4 and $(-1)^{\frac{1}{2}n_1} d(f_1)$ a p-adic square. If $p \neq 2$ this holds for each of the two values 2, 4 of n_1.

Proof. If $p \neq 2$ we use the crude diagonal decomposition of Theorem 31. Without using the full force of the hypothesis $n \geqslant 8$, we permute the variables so that a_{11}, a_{22} and a_{33}, a_{44} are in the same p-adic class; this is certainly possible since there are only four such classes. Then we can take

$$f_1 = a_{11}x_1^2 + a_{22}x_2^2 \quad \text{or} \quad a_{11}x_1^2 + \ldots + a_{44}x_4^2,$$

with

$$(-1)^{\frac{1}{2}n_1} d(f_1) = 4a_{11}a_{22} \quad \text{or} \quad 16a_{11}\ldots a_{44},$$

in either case a p-adic square.

Now let $p = 2$; we coarsen suitably the decomposition (117). If there occur in it unary terms ax_i^2, bx_j^2, $i \neq j$, ab a p-adic square, then we can take $f_1 = ax_1^2 + bx_2^2$ after permuting the variables. Otherwise we take $n_1 = 4$. If in (117) there are five or more unary forms we can find four of them with coefficients $a_1, ..., a_4$ whose product is a 2-adic square, and this as for odd p gives the result.

If there are two products in (117) we can add them together to give a suitable f_1; similarly if there are two terms of the shape $2^r \psi$. We assume therefore (because of $n \geqslant 8$ and what we have just proved) that there is just one product, just one ψ term, and just four unary terms, say $a_i x_i^2$, $i = 1, ..., 4$, with $a_1...a_4$ not a 2-adic square, and no $a_i a_j$ $(i \neq j)$ a 2-adic square. It is fairly easy to deduce that some $a_i a_j$ is in the p-adic class of -1 or of 3, whence $a_i x_i^2 + a_j x_j^2$ added to the ψ or to the product term gives a suitable f_1. This completes the proof; though a little has been left to the reader.

The proof would break down if we required $d(f_1)$ to be in a specified p-adic class other than that of $(-1)^{\frac{1}{2}n_1}$. To see this, suppose that (for odd p)

$$f = x_1^2 + p^2 x_2^2 + ... + p^{2n-2} x_n^2.$$

The theorem, moreover, has no analogue for odd n_1; for if

$$f = c(x_1^2 + p^2 x_2^2 + ...) \quad (p \text{ odd})$$

then with odd n_1 $d(f_1)$ is necessarily in the p-adic class of $\pm c$, according as $n_1 \equiv \pm 1 \pmod 4$.

The object of requiring $d(f_1)$ to be in a specified p-adic class is to enable us later to find a decomposition (119) valid for all p, with some suitably chosen $d(f_1)$ dividing $d(f)$.

7. Quadratic congruences

For given f, p we are interested in the set of integers a for which

$$f(\mathbf{x}) \equiv a \pmod{p^r}, \quad p \nmid \mathbf{x}, \tag{120}$$

is soluble (in integers $x_1, ..., x_n$) for every r. For f can properly represent only these a. It is useful to know that in investigating them we may give r a value depending only on n, d.

THEOREM 37. *For given f, p, a the congruence* (120) *is soluble for all* r:

(i) *if* $p^{n-2} \nmid d$;

(ii) *if* $p^{n-1} \nmid ad$;

(iii) *if* $p^{n-1} \nmid d$ *and* $n \geqslant 5$;

(iv) *if* (120) *is soluble with*

$$r = \begin{cases} u+2+(-1)^p & \text{for} \quad n = 1, 3, \\ u+1 & \text{for} \quad n = 2, \\ u & \text{for} \quad n \geqslant 4, \end{cases} \qquad (121)$$

where $p^u \| d$;

(v) *in case* $a \neq 0$, *if* (120) *is soluble with* r *such that* $p^{r-1} \| 4a$.

Proof. The theorem implies that there exists a least possible integer r_0, dependent only on f, p, such that (120) is soluble for every r if it is soluble for $r = r_0$. Denote this r_0 by $r(f)$ (it depends on p but we suppose p fixed throughout the argument). The assertion $r(f) = 0$ means that (120) is soluble, for the given f, p, for all a and r, since on putting $r = 0$ (120) becomes independent of a. It is clear that

$$r(f) = r(f') \quad \text{if} \quad f \underset{p}{\sim} f'.$$

This formula is to be regarded as asserting that if $r(f')$ exists then $r(f)$ exists and is equal to $r(f')$; and other formulae of the same type, in this proof, are to be interpreted similarly. Plainly we may suppose if convenient that f is equal to the 'standard' form on the right of (105) or (117).

If we partition $\mathbf{x}$ as $\{\mathbf{x}_1, \mathbf{x}_2\}$ then $p \nmid \mathbf{x}$ implies $p \nmid \mathbf{x}_1$ or $p \nmid \mathbf{x}_2$. Hence it is clear that

$$r(f_1+f_2) = 0 \quad \text{if} \quad r(f_1) = 0, \qquad (122)$$

and

$$r(f_1+f_2) \leqslant \max\left(r(f_1), r(f_2)\right). \qquad (123)$$

Another obvious result is

$$r(pf) = 1 + r(f). \qquad (124)$$

(124) means that we get a better upper bound for $r(f) - u$ for forms with $p \,|\, f$, unless $n = 1$; for if $n \neq 1$ we have

$$u(pf) = n + u(f) > 1 + u(f),$$

with an obvious notation. We have also

$$r(x_1x_2) = 0, \quad r(\psi) = 1, \tag{125}$$

with $\psi = \psi_p$ as in (102). $(125)_1$ is trivial. To prove $(125)_2$, see remarks following (104).

If $p^{n-2} \nmid d$, then the right member of (105) or (117) must, when we write $f^{(0)}$ out in full (see (103) and (113)–(115)) begin with a term x_1x_2. Hence (122) and $(125)_1$ give (i). (iv) now follows easily on using (122) and (123), (9). (v) is straightforward; if $\mathbf{x} = \mathbf{y}$ satisfies (120) when $p^{r-1} \| 4a$, then for larger r we put $\mathbf{x} = h\mathbf{y}$ and (120) reduces to a congruence for h which is soluble by the theory of Chapter 3, §1.

In proving (ii) and (iii) we may by (i) suppose $p^{n-2} \| d$, $n \geqslant 2$, whence the hypothesis of (ii) gives $p \nmid a$. With $p^{n-2} \| d$, the right member of (105) or (120) is either $x_1x_2 + \ldots$ (omitted terms involving $x_3, \ldots, x_n$ only) or $\psi + pf_1$, f_1 identically 0 or with $p \nmid d(f_1)$. In the first case (120) is always soluble, in the second it is soluble if $p \nmid a$. This disposes of (ii); and in case (iii) we can suppose $p \mid a$ and apply (i) to the congruence $f_1(x_3, \ldots, x_n) \equiv p^{-1}a \pmod{p^{r-1}}$.

Omitting the restriction $p \nmid \mathbf{x}$ in (120), we have:

THEOREM 38. *Suppose that* $n \geqslant 4$, $a \neq 0$; *then the congruence* $f(\mathbf{x}) \equiv a \pmod{p^r}$ *is soluble for every* r *if it is soluble for* $r = u$, *where* $p^u \| d$.

Proof. We prove the theorem by induction on u; and we note that by Theorem 37 we may suppose $p^2 \mid a$.

Suppose first that $f \equiv 0 \pmod p$, identically. We consider the congruence $p^{-1}f \equiv p^{-1}a \pmod{p^{r-1}}$; in other words we replace d, u, a, r by $p^{-n}d, u-n, p^{-1}a, r-1$. The inductive hypothesis gives what is wanted.

Next suppose $f \equiv a_{11}x_1^2 \pmod p$, $p \nmid a_{11}$ (or more generally, that f is equivalent to a form with this property). Then $f(\mathbf{x}) \equiv a$ $\pmod{p^r}$ is soluble, if at all, with $p \mid x_1$. So we consider the congruence $p^{-1}f(px_1, x_2, \ldots, x_n) \equiv p^{-1}a \pmod{p^{r-1}}$ (the form on the left of which is obviously integral). That is, we replace d, u, r by $p^{2-n}d, u-n+2 < u-1, r-1$. Clearly the inductive hypothesis again gives the result.

Looking at (105), (116) and (117) we see that the only case not dealt with, apart from the obvious $f \underset{p}{\sim} x_1 x_2 + \dots$, is $f \underset{p}{\sim} \psi + pf_1$. So we may suppose $f = \psi + pf_1$. Now since $f \equiv 0$ when and only when $x_1 \equiv x_2 \equiv 0 \pmod{p}$, we have to consider the congruence $p\psi + f_1 \equiv p^{-1}a \pmod{p^{r-1}}$, putting px_1, px_2 for x_1, x_2. Here since we have replaced d, u, r by $p^{4-n}d, u-n+4, r-1$, we see that if $n \geqslant 5$, making $u-n+4 \geqslant r-1$ if $u \geqslant r$, the inductive hypothesis gives the result. If $p \mid f_1$, we can cancel p from the congruence and so replace u, r by $u - 2n + 4 \leqslant u - 4, r - 2$; hence in this case the inductive hypothesis gives the result for $n = 4$ also.

We are left with the case $f = \psi + pf_1$, $n = 4$, $p \nmid f_1$. If f_1 is congruent to a multiple of x_3^2 modulo p (or p-adically equivalent to a form with this property), we can use the inductive hypothesis, considering the congruence

$$\psi + p^{-1}f_1(px_3, x_4) \equiv p^{-2}a \pmod{p^{r-2}}.$$

Otherwise, we have $p \nmid d(f_1)$; and now we show that $f(\mathbf{x}) \equiv a$ $\pmod{p^r}$ is soluble for all a, r. It suffices to prove this for $p \nmid a$ and for $p \| a$. We apply Theorem 37 (ii) to the two congruences $\psi \equiv a$ $\pmod{p^r}, f_1 \equiv p^{-1}a \pmod{p^{r-1}}$ and the proof is complete.

We can obviously improve on Theorem 37 (iv) and Theorem 38 for large n. On the other hand, we can replace $(121)_1$ by $(121)_3$ if $n = 3$, and weaken the hypothesis of Theorem 38 to $n \geqslant 3$, if in case $n = 3$ we impose the further condition that af represents a p-adic square.

8. A property of $\epsilon_p(f)$

We prove a property of $\epsilon_p(f)$ which could be used as the basis of an alternative and in some ways more natural definition of this invariant:

THEOREM 39. *A form f of odd rank n is rationally related to a multiple of a form with discriminant prime to p if and only if $\epsilon_p(f) = 1$.*

Proof. The 'only if' follows from (94), (95) and the rational invariance of $\epsilon_p(f)$. So we assume $\epsilon_p(f) = 1$ and prove the 'if'. We may suppose that d is divisible exactly by an even power of p; for if not, consider in place of f the form df, with discriminant d^{n+1}, which is a square since n is odd.

We may also suppose that f does not satisfy the condition

$$p \,|\, a_{1n}, \ldots, a_{n-1,n}, p^2 \,|\, a_{nn}. \tag{126}$$

For if so, we may consider instead of f the form

$$f(x_1, \ldots, x_{n-1}, p^{-1}x_n),$$

which is integral and has discriminant $p^{-2}d$. More generally, we may assume that there is no integral $\mathbf{y}$ with

$$p \,|\, A\mathbf{y}, \quad p^2 \,|\, f(\mathbf{y}), \quad p \nmid \mathbf{y}. \tag{127}$$

For obviously (127) is soluble, if at all, with $\mathbf{y}$ primitive; and if so, f is equivalent to a form satisfying (126), which is just (127) with $\mathbf{y} = \{0, 1\}$. But the insolubility of (127) is clearly invariant under p-adic equivalence. We may therefore suppose that the right member of (105) or (117) does not satisfy (126) or (127). The proof will be complete if with this further assumption we can prove $p \nmid d$.

Now the right member of (105) or (117) satisfies (126) unless it reduces to $f^{(0)} + pf^{(1)}$. With this, it still satisfies (126), after an obvious equivalence transformation (a permutation of the variables), if $f^{(1)}$ contains a product term, as it must if

$$n(f^{(1)}) = n_1 > 2.$$

So we suppose $n_1 \leqslant 2$. If $n_1 = 0$, then $f \underset{p}{\sim} f^{(0)}$, which must since $n = n_0$ is odd be $\sigma + dx_n^2$, $p \nmid d$. If $n_1 = 1$, then an odd power of p (possibly 8 if $p = 2$) divides d exactly, contrary to hypothesis. So we assume $n_1 = 2$. Using (116) in case $p = 2$, we see that this gives us (126) unless $f^{(1)} = \psi$. We complete the proof by assuming this and showing that $\epsilon_p(f) = -1$, contrary to hypothesis. We have $p \nmid d(f^{(0)})$, $d(\psi)$, whence $\epsilon_p(f^{(0)}) = \epsilon_p(\psi) = 1$ by (95); but $\epsilon_p(p\psi) = (p, d(\psi))_p$ by (94), $= -1$ by (78). So

$$\epsilon_p(f) = \epsilon_p(f^{(0)} + p\psi) = -(d(f^{(0)}), -p^2 d(\psi))_p$$

by (93). Using (77), (79), (80) and (52) if $p = 2$, we obtain the desired contradiction $\epsilon_p(f) = -1$ and complete the proof.

If we used this result as the definition of $\epsilon_p(f)$ for $p \neq \infty$ and n odd, we could deal with the case of even n by saying that (see

(94)) $\epsilon_p(f) = \epsilon_p(f - x_{n+1}^2)$ for n even. And for $p = \infty$ we could fall back on (91).

The theorem remains valid for even n if the condition $\epsilon_p(f) = 1$ is replaced by (i) u even, where $p^u \| d$, and (ii) *either* $\epsilon_p(f) = 1$ *or* d not a p-adic square. (The parity of u is rationally invariant always, and unaffected by removing a divisor when n is even.)

Write temporarily $\phi = dx_{n+1}^2$ or $x_{n+1}^2 - dx_{n+2}^2$ according as n is odd or even. Then it is easily seen that $\epsilon_p(f) = 1$ if and only if $f - \phi$ is rationally related to a disjoint sum of binary p-adic zero forms, for $p \neq \infty$.

CHAPTER 5

THE CONGRUENCE CLASS AND THE GENUS

1. Congruential equivalence

We shall say that two forms f, f' are CONGRUENTIALLY EQUI-VALENT, or are in the same CONGRUENCE CLASS, in symbols $f \cong f'$, if for every $m \neq 0$ there exists a third form f'', depending on m, such that

$$f'' \sim f, \quad f'' \equiv f' \pmod{m}, \tag{128}$$

identically in the variables. It is not difficult to show directly that this relation has the properties (16)–(18) of Chapter 1; this follows also from Theorem 41, below.

THEOREM 40. *The rank and discriminant, and all the $\epsilon_p(f)$, are invariant under congruential equivalence.*

Proof. For the invariance of n, d, take $m = 2|d(f)| + 2|d(f')|$ in (128). For that of $\epsilon_p, p = 2, 3, 5, \ldots$, take m to be a prime power, and compare (97) and (128). Now (92) gives the invariance of ϵ_∞.

The relation between congruential and p-adic equivalence is given by

THEOREM 41. *Two forms f, f' of the same rank n and the same discriminant d are congruentially equivalent if and only if $f \underset{p}{\sim} f'$ for $p = 2, 3, 5, \ldots$.*

Proof. The 'only if' is clear on comparing (97) and (128). Now assume $f \underset{p}{\sim} f'$; we first show, using $d(f) = d(f') = d \ (\neq 0)$ that (97) can be satisfied with $|P| \equiv \pm 1 \pmod{p^t}$, for every t. To see this, consider (97) with $t + k$ for t, where $p^k \| 2d$. This gives $|P|^2 d \equiv d, d(|P| - 1)(|P| + 1) \equiv 0 \pmod{p^{t+k}}$. As the g.c.d. of the numbers $|P| \pm 1$ is at most 2, this gives $|P| \equiv \pm 1 \pmod{p^t}$.

The next step is to show that for all f, p^t there is an integral U with $|U| \equiv -1 \pmod{p^t}$ such that $f^U \equiv f \pmod{p^t}$. This property is trivial for the forms on the right of (105) and (117) (the substitution $\mathbf{x} \to U\mathbf{x}$ may be taken to be $x_1 \to -x_1$ or interchange of x_1, x_2). To prove it generally, note that if f^P has it, with $U = V$, say, where $p \nmid |P|$, then f has it with $U = PVQ$, with Q chosen as in the remark following (97).

Combining these two results, for p^t ranging over the prime power factors of $m \neq 0$, we see that we may take $|P| \equiv +1$ in the first and so have $f^M \equiv f'$ (mod m) for some integral M with $|M| \equiv \pm 1$ (mod m). It suffices to find an integral T with $|T| = \pm 1$ and $T \equiv M$ (mod m), and take $f'' = f^T$ in (128).

For $n = 1$ this is trivial, so we take $n \geqslant 2$ and use induction. We may suppose, by adding suitable multiples of m to its elements, that $\mathbf{z}$, the first column vector of M, is primitive. Using Theorem 1, construct an integral Z with $|Z| = \pm 1$ and first column $\mathbf{z}$. Express M as ZN, where N clearly has first column $\{1, \mathbf{0}\}$ and so may be written as

$$\begin{pmatrix} 1 & \mathbf{y}' \\ 0 & N_1 \end{pmatrix}, \quad \text{with} \quad |N_1| = |Z|^{-1}|M| \equiv \pm 1 \;(\text{mod } m).$$

It suffices to take

$$T = Z \begin{pmatrix} 1 & \mathbf{y}' \\ 0 & T_1 \end{pmatrix}, \quad \text{with } T_1 \equiv N_1 \;(\text{mod } m),$$

whence the inductive argument is clear. (We could of course take $|T| = |M| = +1$.)

COROLLARY. *The condition* $f \underset{p}{\sim} f'$ $(p = 2, 3, 5, \ldots)$ *of Theorem* 41 *may be weakened to* $f \underset{p}{\sim} f'$ *for* $p \,|\, d$, *or to* (128) *with* $m = d$.

Proof. Theorems 29 and 30, 33.

2. Construction of congruence class with prescribed properties

Theorem 41 shows that the congruence class of a form f is determined uniquely if we know that, for given $n \geqslant 1$, $d \neq 0$, and $f_2, f_3, \ldots, f$ satisfies

$$n(f) = n, \quad d(f) = d, \quad f \underset{p}{\sim} f_p \quad (p = 2, 3, \ldots). \tag{129}$$

Clearly every f satisfies (129) with suitable n, d, f_p; each f_p may (but need not) be taken to be one of the 'standard' forms on the right of (105) or (117). We consider the problem of constructing f to satisfy (129), with given n, d, f_p; we first find some necessary conditions on n, d, and the f_p, then prove them sufficient.

First we must have

$$n(f_p) = n, \quad dd_p \text{ a } p\text{-adic square} \quad (p = 2, 3, \ldots), \quad (130)$$

where $d_p = d(f_p)$. Next we must have

$$p^u \| d_p, \quad \text{for } u = u(p) \text{ such that } p^u \| d \quad (p = 2, 3, \ldots). \quad (131)$$

The necessity of these two conditions follows immediately from the definition of p-adic equivalence and $(129)_3$. We shall see that in general these two conditions are sufficient; but we must have also

$$\prod_{p=2, 3, \ldots} \epsilon_p(f_p) = 1 \quad \text{in case } n = 2 \text{ and } d > 0. \quad (132)$$

To see the necessity of (132), note that $(129)_3$ and Theorem 28 give us $\epsilon_p(f) = \epsilon_p(f_p)\,(p = 2, 3, \ldots)$, whence the left member of (132) is $\epsilon_\infty(f)$ by (92). Now when $n = 2$ and $d > 0$ we have $s = 0$ and $\epsilon_\infty(f) = 1$ by (91).

Write for brevity
$$a_p = f_p(1, 0, \ldots, 0).$$

There is no loss of generality in supposing $a_p \neq 0$. Then if $n = 2$ we have $\epsilon_p(f_p) = (a_p, d_p)_p$ by (98), $= (a_p, d)_p$ by (77) and (130), so (132) may be rewritten as

$$\prod_{p=2, 3, \ldots} (a_p, d)_p = 1 \quad \text{if} \quad n = 2 \text{ and } d > 0. \quad (133)$$

The construction of f satisfying (129) would be easier if all the a_p were equal. Let us consider therefore, for given a, p, whether there exists f_p' with leading coefficient a and with $f_p' \underset{p}{\sim} f_p$. If there is, then putting f_p' for f_p does not affect any of (129)–(132). It is clear from Theorem 37 that a sufficient condition on a for the existence of such an f_p' is $a \equiv a_p \pmod{p^{u+3}}$. This condition can obviously be satisfied for any finite number of p simultaneously, with a either positive or negative as we choose. Consider therefore the case $p \nmid d$. In this case, if also $n \geqslant 3$, no restriction on a is needed; for Theorems 29 and 30 show that (for some $(n-2)$-ary g)

$$f_p \underset{p}{\sim} x_1 x_2 + g \underset{p}{\sim} a x_1^2 + x_1 x_2 + g,$$

for any a (put $x_2 + a x_1$ for x_2). Similarly, if $p \nmid d$, $n = 2$, and d is a p-adic square we need not restrict a; for we have

$$f_p \underset{p}{\sim} x_1 x_2 \underset{p}{\sim} a x_1^2 + x_1 x_2.$$

Now suppose $n = 2$, $p \nmid d$, and d not a p-adic square. It suffices to impose the restriction $p \nmid a$. For Theorem 30 and (104), with $h = a \not\equiv 0 \pmod{p}$, give $f_p \underset{p}{\sim} \psi_p \underset{p}{\sim} a\psi_p$, ψ_p, see (102), having leading coefficient 1.

Using these results and varying p, we prove:

LEMMA. *Conditions* (130)–(132) *imply the existence of an integer* $a \neq 0$, *which may be chosen either positive or negative if* $n \geq 3$, *such that each* f_p *is* p-*adically equivalent to a form* f_p' *with leading coefficient* a.

Proof. For $n = 1$ take $a = d$. For $n \geq 3$, the result is clear from the foregoing remarks.

Assume therefore $n = 2$. From what has been said it is clear that there is an $a = a'q$ such that $q \nmid 2d$ is prime, while a' has no prime factor not dividing d, which does what is wanted for $p \mid d$, and therefore for $p \neq q$. (We here appeal to Dirichlet's theorem on primes in an arithmetical progression.) Moreover, we are free to choose either a positive or a negative a'. If we can prove that (for suitable choice of sgn a') d is a q-adic square, then we have what is wanted for $p = q$ also. Now $(d \mid q) = (a'q, d)_q = (a, d)_q$ by (78) and the restrictions imposed on a. So it suffices to prove $(a, d)_q = 1$.

It is clear that $(a, d)_p = (a_p, d)_p$ for $p \neq \infty$, q, each of these symbols being $\epsilon_p(f_p) = \epsilon_p(f_p')$. From (83) we have therefore $(a, d)_\infty (a, d)_q = \epsilon$, where ϵ denotes the left member of (132) or (133), and is 1 if $d > 0$. In case $d > 0$ we have therefore $(a, d)_q = 1$ as required, while if $d < 0$ we have $(a, d)_q = \epsilon \operatorname{sgn} a$, and we are free to choose sgn $a = \epsilon$.

The lemma gives us, completing the square,

$$4af_p \underset{p}{\sim} (2ax_1 + l_p)^2 + g_p \quad (p = 2, 3, \dots), \tag{134}$$

for certain linear forms l_p and quadratic g_p, in $x_2, \dots, x_n$, in case $n \geq 2$. We now state formally:

THEOREM 42. *Conditions* (130)–(132) *are necessary and sufficient for the existence of* f *satisfying* (129).

Proof. We have already noted the necessity of the conditions. Their sufficiency is trivial for $n = 1$ (take $f = dx_1^2$); so we assume $n \geq 2$ and proceed by induction.

It is easy to see from (130), (131), (134) and (42) that the forms g_p satisfy (130) and (131) with $n-1$, d_1, g_p for n, d, f_p, where

$$d_1 = (-1)^{n-1}(16)^{[\frac{1}{2}n-\frac{1}{2}]}a^{n-2}d.$$

If $n = 3$, we choose sgn a so that $d_1 = 16ad < 0$, and the condition on the g_p corresponding to (132) becomes vacuous. There is therefore a form g satisfying (129) with $n-1$, d_1, g, g_p for n, d, f, f_p, by the inductive hypothesis. Hence from (134) we have

$$4af_p \underset{p}{\sim} (2ax_1+l'_p)^2 + g(x_2..., x_n) \quad (p = 2, 3, ...),$$

for some linear forms l'_p. We can choose a linear form l with $l \equiv l'_p \pmod{p^v}$, for each prime power p^v with $p^v \parallel 2a$.

Now we have (for $p = 2, 3, ...$)

$$4af_p \underset{p}{\sim} (2ax_1+l)^2 + g(x_2, ..., x_n).$$

The form on the right is clearly expressible as $4af$, where f is integral and satisfies $(129)_1$ and $(129)_3$; and using (42) again we see that also $(129)_2$ holds, that is, $d(f) = d$; whence the proof is complete.

COROLLARY. *There is an f satisfying $(129)_1$ and $(129)_3$ if and only if (130) and (132) are satisfied for some d, and in addition we have $p \mid d_p = d(f_p)$ for at most finitely many p.*

Proof. If the hypotheses of the corollary hold with $d = d'$ then those of the theorem hold with $d = d''$, for some d'' such that $d' d''$ is a perfect square; and conversely.

3. Semi-equivalence

Two forms f, f' are said to be SEMI-EQUIVALENT, in symbols $f \simeq f'$, if they are congruentially equivalent and also have the same signature. It is clear that this relation has the usual properties (16)–(18) discussed in Chapter 1. The family of forms f' semi-equivalent to a given f is called the GENUS of f.

THEOREM 43. *The genus of a form f is uniquely determined if it is given that f satisfies*
$$f \simeq f_0, \quad s(f) = s, \tag{135}$$
for a given f_0 and a given integer s. The necessary and sufficient conditions for the existence of f satisfying (135) are
$$|s| \leqslant n = n(f_0), \quad s \equiv s(f_0) \pmod 8. \tag{136}$$

Proof. The first assertion follows immediately from the definition. Conversely, every f satisfies (135) for some f_0, s; we can take $f_0 = f$, or suppose (by the second part of the theorem) that $-3 \leqslant s(f_0) \leqslant 4$.

The necessity of (136)$_1$ is trivial. For the necessity of (136)$_2$, note that by Theorem 40 and the definition of semi-equivalence (135) implies

$$n(f) = n(f_0), \quad d(f) = d(f_0), \quad \epsilon_\infty(f) = \epsilon_\infty(f_0).$$

Using these and (23) and (91), we deduce

$$s(f) \equiv s(f_0) \equiv n \ (\text{mod } 2), \quad [\tfrac{1}{2}s(f)] \equiv [\tfrac{1}{2}s(f_0)] \ (\text{mod } 2),$$

whence
$$s(f) \equiv s(f_0) \ (\text{mod } 4),$$

and finally
$$[\tfrac{1}{4}s(f) + \tfrac{1}{4}] \equiv [\tfrac{1}{4}s(f_0) + \tfrac{1}{4}] \ (\text{mod } 2),$$

giving (136)$_2$.

The congruence class and the genus coincide if $n \leqslant 3$; for then (136) gives $s = s(f_0)$ (and the sufficiency follows trivially, taking $f = f_0$). The integer a used in the proof of Theorem 42 can be chosen either positive or negative as we please if $n \geqslant 4$; hence the sufficiency of (136) can be proved for $n \geqslant 4$ by an argument like that of Theorem 27.

The case $n = 2$ can be dealt with alternatively as follows:

THEOREM 44. *Suppose $d \neq 0$, $d \equiv 0$ or 1 (mod 4), and g.c.d. $(a, d) = 1$. Then* (i) *a binary form f such that $f \equiv a$ (mod d) is soluble exists if and only if either $d < 0$ or $\prod\limits_{p \mid d} (a, d)_p = 1$, and* (ii) *if such an f exists its genus is uniquely determined.*

Proof. We first show that if there does exist an f with the stated properties, then $f \equiv a$ (mod p^t) is soluble for each p dividing d, and all t. This is immediate except for $p = 2$, $d \equiv 4$ (mod 8), in which case we use Theorem 34. It follows easily from Theorems 32 and 35 that for $p \mid d$ we have $f \underset{p}{\sim} f_p$, where $f_p = ax_1^2 - adx_2^2$ for odd d, $ax_1^2 - \tfrac{1}{4}adx_2^2$ for even d. By Theorem 41, Corollary, this determines the congruence class, which for $n < 4$ coincides with the genus, uniquely. This disposes of (ii). Defining f_p to be $x_1 x_2$ or ψ_p (see (102)) when $p \nmid d$, the case $d < 0$ of (i) follows from

Theorem 42. In case $d > 0$, condition (133) reduces to

$$\prod_{p|d} (a, d)_p = 1,$$

and this completes the proof.

4. Decomposition under congruential equivalence and semi-equivalence

Given a form f, of rank n and discriminant d, we seek a decomposition of the shape

$$f \simeq f_1 + f_2, \qquad (137)$$

in which the forms f_1, f_2 have prescribed ranks n_1, n_2 and

$$d_1 = d(f_1), \quad d_2 = d(f_2)$$

are such that

δd_1 is a square, whence $(-1)^{n_1 n_2} \delta d d_2$ is a square,   (138)

by (9), for some given $\delta \neq 0$. An obviously necessary condition for the existence of such a decomposition is that for $p = 2, 3, \ldots$ there must exist decompositions

$$f \underset{p}{\sim} f_1^{(p)} + f_2^{(p)}, \quad n(f_1^{(p)}) = n_1, \quad \delta d(f_1^{(p)}) \text{ a } p\text{-adic square.} \quad (139)$$

For if we put $f_1^{(p)}$, $f_2^{(p)} = f_1$, f_2, for all p, then (137) and (138) imply (139).

Conversely, we have:

THEOREM 45. *The conditions* (139), *for* $p = 2, 3, \ldots$, *are sufficient for the existence of a decomposition of the shape* (137) *satisfying* (138), *provided* $\delta < 0$ *if* $n_1 = 2$, *and* $\delta d < 0$ *if* $n_2 = 2$; *and if* $n \geqslant 8$ *they are satisfied with* $n_1 = 2$ *or* 4 *and* $\delta = (-1)^{\frac{1}{2} n_1}$.

Proof. Suppose that (139) holds for all p. Then except for $p \mid d$ we must have $p \nmid d(f_1^{(p)})$. So by the corollary to Theorem 42 there exists an f_1 (independent of p) with $f_1 \underset{p}{\sim} f_1^{(p)}$ for all p. With f_2 similarly constructed, (138) follows from (139) and we have $f \underset{p}{\sim} f_1 + f_2$ for all p, giving (137), by Theorem 41. This proves the first assertion. For the second, see Theorem 36.

We deduce:

THEOREM 46. *Let f be a form with rank $n \geqslant 8$, and discriminant $d \neq 0$. Then we have*

$$f \simeq f_1 + \ldots + f_k \quad (k \geqslant 2), \qquad (140)$$

for some forms f_i such that $n_k = n(f_k) \leqslant 7$ and

$$\left.\begin{aligned} n_i &= n(f_i) = 2 \text{ or } 4, \\ \text{and } (-1)^{\frac{1}{2}n_i} & d(f_i) \text{ is a square} \end{aligned}\right\} \text{ for } i = 1, ..., k-1. \quad (141)$$

The signatures s_i of the forms f_i satisfy

$$s_i \equiv n_i \pmod 4 \text{ for } i = 1, ..., k-1, \quad (142)$$

$$s_1 + ... + s_k \equiv s(f) \pmod 8. \quad (143)$$

Proof. Repeated application of Theorem 45 gives the decomposition (140) satisfying (141), whence (142) follows by (23), and (143) by Theorem 43. The process stops when after splitting off $f_1, ..., f_{k-1}$ we are left with an f_k of rank less than 8.

We now consider whether we can convert (137) into a decomposition under semi-equivalence by modifying it in a way which is trivial under the wider relation of congruential equivalence. By the definition of semi-equivalence we have (assuming (137))

$$f \simeq \phi_1 + \phi_2$$

if ϕ_1, ϕ_2 satisfy

$$\phi_1 \cong f_1, \quad \phi_2 \cong f_2, \quad s(\phi_1) + s(\phi_2) = s(f).$$

By Theorem 43, ϕ_1, ϕ_2 can be found to satisfy these conditions, and to have signatures s_1', s_2', if there exist s_1', s_2' such that

$$|s_1'| \leqslant n_1, \quad |s_2'| \leqslant n_2, \quad s_1' \equiv s_1 \pmod 8, \quad s_2' \equiv s_2 \pmod 8,$$

and

$$s_1' + s_2' = s = s(f). \quad (144)$$

It may be noted that the fourth of conditions (144) is redundant since (cf. (143)) we have $s_1 + s_2 \equiv s \pmod 8$. If we assume (cf. (142)) that $s_1 \equiv n_1 \equiv -n_1 \pmod 4$, then it is clear that (144) can always be satisfied if $|s| \leqslant n-4$, but cannot be satisfied in the cases

$$s_1 \equiv n_1 - 4 \pmod 8, \quad s \geqslant n-2,$$

$$s_1 \equiv 4 - n_1 \pmod 8, \quad s \leqslant 2-n.$$

THEOREM 47. *Suppose that $n = n(f) \geqslant 8$, and in case*

$$|s| = |s(f)| \geqslant n-2,$$

make the stronger assumption $n \geqslant 12$. Then f has a decomposition

$$f \simeq \phi_1 + ... + \phi_v \quad (v \geqslant 2),$$

in which each form ϕ_i with $i < v$ is of even rank not exceeding 8, *while ϕ_v is of rank at most* 11 *in any case, and at most* 7 *if* $|s| \leqslant n-4$.

Proof. We consider the decomposition (140), and first suppose that the left member of (143) is less than the right. We bracket in pairs those of the f_i $(i = 1, ..., k-1)$ which have $s_i \equiv n_i - 4 \pmod 8$; if one such f_i is left over we bracket it with f_k if $s \geqslant n-2$. The rest of the argument will be clear from the foregoing remarks. The case in which the left member of (143) is the greater is dealt with similarly; and in the remaining case (140) holds with $\simeq$ for $\cong$.

5. Rational relations between semi-equivalent forms

From the fact that n, d, s and all the $\epsilon_p(f)$ are invariants under semi-equivalence it follows that forms in the same genus are rationally related. We shall prove a stronger result than this. To do so we need to know a little more about the solubility of Diophantine equations. The result needed, which will be useful again later, is less cumbrous when expressed in a shape not invariant under equivalence:

THEOREM 48. *Suppose that $n \geqslant 2$, af is either indefinite or positive, $adm_1 \neq 0$, and $a_{11} \equiv a \pmod{4a^2 dm_1}$. Then there exist integral $\mathbf{x}$, h such that*

$$f(\mathbf{x}) = ah^2, \quad \mathbf{x} \equiv h\{1, \mathbf{0}\} \pmod{adm_1},$$
$$\text{g.c.d. } (h, adm_1) = 1. \quad (145)$$

Proof. We consider the form $f(\mathbf{x}) - ax_{n+1}^2$, $= \phi$, say. Since af is indefinite or positive, ϕ is indefinite. The congruence condition imposed on a_{11} gives us, by Theorem 37, that $f \equiv a \pmod m$ is soluble for every $m \neq 0$. Clearly this makes ϕ a p-adic zero form for $p = 2, 3, ...$; so by Theorems 14 and 15, ϕ is a rational zero form. We may therefore appeal to Theorem 16, with $\phi, n+1 \geqslant 3$, $\{1, 0, ..., 0, 1\}$, adm_1 for $f, n, \mathbf{t}, q$. With these substitutions (58) reduces to $f(\mathbf{x}) = ax_{n+1}^2$ with $x_{n+1} \equiv h \pmod{adm_1}$ and $\mathbf{x}, h$ satisfying $(145)_2$ and $(145)_3$. The last two formulae are unaffected on putting $h = x_{n+1}$, which gives us $(145)_1$.

It suffices therefore to verify that the hypothesis (57) of

Theorem 16 is satisfied. This clearly follows if we show that, for every $m \neq 0$, there exists an integral vector $\mathbf{y}$ with

$$f(\mathbf{y}) \equiv a \pmod{m}, \quad \mathbf{y} \equiv \{1, 0\} \pmod{adm_1}. \qquad (146)$$

Now we may take m to be a power of a prime p. If $p \nmid adm_1$ we have to consider $(146)_1$ only, and have already noted that it is soluble. So we assume $p \mid adm_1$. We put $\mathbf{y} = k\{1, 0\}$ in (146) and so see that it is sufficient to find an integer k such that

$$k^2 a' \equiv 1 \pmod{p^t}, \quad k \equiv 1 \pmod{adm_1},$$

where $a' = a^{-1} a_{11} \equiv 1 \pmod{4adm_1}$, for any given $p \mid adm_1$ and $t \geqslant 1$. Plainly $a' \equiv 1 \pmod{p}$ always, and $a' \equiv 1 \pmod{8}$ if $p = 2$; so the congruence for k^2 is certainly soluble. It is easy to see that one of the solutions has $k \equiv 1 \pmod{p^u}$, where $p^u \| adm_1$; for $p^u \mid (k+1)(k-1)$ always, and $2^{u+1} \mid (k+1)(k-1)$ if $p = 2$, while the g.c.d. of $k \pm 1$ is prime to p if $p \neq 2$, oddly even for $p = 2$. This completes the proof.

THEOREM 49. *With the hypotheses of Theorem 48, and with h as in (145), there is a rational R such that $|R| = \pm 1$, hR is integral, $R \equiv I \pmod{adm_1}$, and f^R is an integral form with leading coefficient a and identically congruent to f modulo adm_1, provided $|adm_1| \geqslant 2$.*

Proof. With $adm_1 \neq 0, \pm 1$, (145) shows that $h \neq 0$, and the g.c.d. of $x_1, \ldots, x_n$ divides h. We may therefore suppose $\mathbf{x}$ primitive, otherwise (145) would hold with a smaller h. Now there is an integral T (see Theorem 1) with $\mathbf{x}$ as its first column and with $|T| = \pm 1$. f^T has ah^2 as its leading coefficient; and we may suppose $T \equiv [h, M] \pmod{adm_1}$ for some M (if not, add suitable multiples of the first column to the others).

Next (see Theorem 12) there is an integral T_1, $(n-1)$ by $(n-1)$, with determinant ± 1, such that $f^{T[1, T_1]}$, which also has leading coefficient ah^2, has no terms in $x_1 x_j$, $j > 2$. This means that the substitution $\mathbf{x} \to T[1, T_1][h^{-1}, h, 1, \ldots, 1]\mathbf{x}$ takes f into an integral form.

We shall now take $R = T[1, T_1][h^{-1}, h, 1, \ldots, 1][1, T_2^{-1}]$, with T_2 integral, $|T_2| = \pm 1$. Plainly this makes hR integral and we have only to show that by suitable choice of T_2 it makes $R \equiv I$

(mod adm_1). This congruence is to be understood to mean $hR \equiv hI$ (mod adm_1), and can be expressed as

$$T[1, T_1][h^{-1}, h, 1, \ldots, 1][1, T_2^{-1}] \equiv I \pmod{adm_1}.$$

Since we chose $T \equiv [h, M]$ (mod adm_1), with an M clearly satisfying $h\,|M| \equiv \pm 1$ (mod adm_1), this congruence reduces to

$$T_2 \equiv MT_1[h, 1, \ldots, 1] \pmod{adm_1}.$$

So by an argument used in the proof of Theorem 41 it can be satisfied, since

$$|MT_1[h, 1, \ldots, 1]| \equiv \pm |T_1| (= \pm 1) \pmod{adm_1}.$$

We deduce a well-known classical result:

THEOREM 50. *Two integral quadratic forms are in the same genus if and only if they are related by a rational transformation whose matrix has determinant* ± 1 *and denominator prime to their discriminant. If so, then the rational transformation may be so chosen that the denominator of its matrix is prime to any prescribed non-zero integer.*

Proof. To prove the 'if', suppose f, f' satisfy $f^R = f'$ identically, with $|R| = \pm 1$ and hR integral for an h prime to $d = d(f)$, where R, h have not necessarily the values of Theorem 49. Then obviously $d(f') = |R|^2 d = d$, and $s(f') = s(f)$; so by the corollary to Theorem 41 it suffices to prove $f \underset{p}{\sim} f'$ for $p\,|\,d$, implying $p \nmid h$. Now for such p, and any t, we can clearly satisfy (97) by choosing an integral P with $hP \equiv hR$ (mod p^t).

We prove the other two assertions simultaneously by induction on n. They are both trivial for $n = 1$ since with $n = 1$, $f \simeq f'$ implies $f = f' = dx_1^2$, where $d = d(f) = d(f')$. It is sufficient to consider primitive forms; we may therefore suppose, by a preliminary equivalence transformation, that one of the given forms, say f', has leading coefficient $a \neq 0$, prime to m, where $m \neq 0$ is the given integer, required to be relatively prime to the denominator of the transformation. Clearly this makes af' either indefinite or positive; whence also af is either indefinite or positive, f being the other given form. By an equivalence transformation on f, we may suppose $a_{11} \equiv a$ (mod $4a^2 dm_1$), where $m_1 \neq 0$ is to be chosen suitably (so as to be a multiple of m).

Now using Theorem 49 we may suppose (with R as in that theorem)

$$4af^R = (2ax_1+l)^2+g, \quad 4af' = (2ax_1+l')^2+g', \quad (147)$$

where l, l' are linear and g, g' quadratic in $x_2, ..., x_n$. Plainly $d(g) = d(g')$ and $s(g) = s(g')$, and $g \equiv g' \pmod{4a^2dm_1}$. Choosing m_1 suitably, this congruence gives $g \cong g'$, by the corollary to Theorem 41. Now with $s(g) = s(g')$ we have $g \simeq g'$, so by the inductive hypothesis there exists an $(n-1) \times (n-1)$ matrix Q, with $|Q| = \pm 1$, and with denominator prime to $2m$, such that $g^Q = g'$.

Now f is taken into f' by the substitution $\mathbf{x} \to R[1,Q] V\mathbf{x}$, where V is the matrix of the substitution $x_1 \to x_1 + (2a)^{-1}(l'-l^Q)$. By our choice of a this gives what is wanted if we can prove $l' \equiv l^Q \pmod 2$. (l^Q is the transform of l by Q). But (147) gives

$$l^2 \equiv g, \quad (l^Q)^2 \equiv g^Q = g' \equiv (l')^2 \pmod 2,$$

and this completes the proof.

6. Representation of integers

If f represents $a \neq 0$ then af represents $a^2 > 0$ and so

$$af \text{ is either indefinite or positive.} \quad (148)$$

Another necessary condition for a representation of a by f is the solubility, for every $m \neq 0$, of

$$f(\mathbf{x}) \equiv a \pmod m. \quad (149)$$

The representation cannot be proper unless the solution of (149) can be chosen to be primitive, or equivalently to satisfy

$$\text{g.c.d. } (x_1, ..., x_n, m) = 1. \quad (150)$$

These necessary conditions, which may be called the GENERIC CONDITIONS for a representation (or for a proper representation) of a by f, are not sufficient. For a simple example, take

$$f = 2x_1^2 + x_1x_2 + 3x_2^2, \quad a = 1.$$

Clearly f has minimum 2 and does not represent 1, yet (149) is always soluble, as the next theorem shows. An important property of the genus is that if we consider representations of a by forms in the genus of f then these conditions, while remaining necessary, become sufficient.

THEOREM 51. (i) *Suppose that* (148) *holds and that for every* $m \neq 0$ *there are integers* x_i *satisfying* (149). *Then some form* f' *in the genus of* f *represents* a. *Moreover, it suffices to take* $m = 4a^2d$ *if* $n = 2$ *or* 3, $m = d$ *if* $n \geqslant 4$.

(ii) *The form* f' *can be chosen so as to represent* a *properly if the solution of* (149) *can for every* $m \neq 0$ *be chosen to satisfy* (150). *When the restriction* (150) *is imposed, it suffices to take* $m = ad$ *if* $n = 2$, d *if* $n \geqslant 4$, $4d \prod_{p \mid d} p$ *if* $n = 3$.

Proof. We begin by remarking that the conditions are necessary; for the definition of semi-equivalence shows that they are unaltered on putting f' for f. That they can be weakened for $n \geqslant 2$, as asserted, by giving m a value depending only on a, d, is easily deduced from Theorems 37 and 38. We therefore prove them sufficient in their stronger shapes.

If $a_{11} \equiv a \pmod{8a^2d}$ then Theorem 49 (with $m_1 = 2$) enables us to construct a form $f' = f^R$ with leading coefficient a, which by Theorem 50 is in the genus of f. By Theorem 1, it follows that the solubility of (149) and (150) with $m = 8a^2d$ implies the conclusion of part (ii) of this theorem. To deduce (i), note that the hypotheses of (i), with $a = a_1 a_2^2$, a_1 square-free, hold if and only if those of (ii) hold for $a = a_1 a_3^2$, for some $a_3 \mid a_2$.

COROLLARY. *With the hypotheses of part* (ii) *of Theorem* 51, f *itself represents properly, for every* $m_1 \neq 0$, *an integer* ah^2 *with* h *prime to* m_1.

Proof. As for the theorem, but appealing to Theorem 48 instead of Theorem 49.

The obvious analogue of this corollary, for representation not restricted to be proper, can be sharpened:

THEOREM 52. *Suppose* $n \geqslant 3$, $m_1 \neq 0$; *then there is an integer* $k \neq 0$, *prime to* m_1, *depending only on* f *and* m_1, *and such that* f *represents every* a *satisfying, besides the conditions of Theorem* 51 (i), *the further condition* $a \equiv 0 \pmod{k^2}$.

Proof. There are by Theorem 11 only finitely many classes of forms in the genus of f. Hence the f' of Theorem 51 (i) can be taken to be one of a finite set of forms $f_1, f_2, \ldots$. These forms can, by Theorem 50, be obtained from f by transformations with denominators, say $h_1, h_2, \ldots$, all prime to m_1. Take k to be the least common multiple of $h_1, h_2, \ldots$; then clearly k is prime to m_1, and f' in Theorem 51 (i) is always expressible as f^R, with rational R such that kR is integral. And we may suppose k prime to d also.

We now observe that (148) is unaffected, while (149) remains soluble for every $m \neq 0$, if we replace $a \equiv 0 \pmod{k^2}$ by ak^{-2}. The first assertion is trivial. If, as we may, we take m to be a prime power p^t, the second assertion is trivial except for $p \mid k$, implying $p \nmid d$; and then it is trivial by Theorems 29 and 30, and $n \geqslant 3$.

Now appealing to Theorem 51 (i) we have $f(R\mathbf{x}) = ak^{-2}$, $f(kR\mathbf{x}) = a$, for integral $\mathbf{x}$ and R such that kR is integral. Hence $kR\mathbf{x}$ is integral and the proof is complete.

Theorem 52 enables us to construct arithmetic progressions, every integer in which is represented by a given form f. This introduces a useful element of linearity into the problem of representation of integers; we take advantage of this in the next section and obtain a much stronger result.

7. Representation by indefinite forms with $n \geqslant 4$

We show now that the generic conditions of the last section are sufficient for a representation or proper representation of a by f if $n \geqslant 4$ and f is indefinite. It suffices to consider proper representation.

THEOREM 53. *Suppose that $n \geqslant 4$, f is indefinite, and f represents properly an integer congruent to a modulo d. Then f represents a properly.*

Proof. The case $a = 0$ has been dealt with in Chapter 3, §5; so we assume $a \neq 0$. Now by Theorem 51 (ii) we see that the congruence condition of the present theorem implies the solubility of $f(\mathbf{x}) \equiv a \pmod{m}$, with primitive $\mathbf{x}$, for every $m \neq 0$. Take $m = 8a^2d$; appealing to Theorem 1, we see that by a preliminary equivalence transformation we may suppose (temporarily) $a_{11} \equiv a \pmod{8a^2d}$. Then Theorem 48 shows that (145) holds with $m_1 = 2$, for some h prime to $2ad$ and some $\mathbf{x}$ which may be taken to be primitive by choosing a smaller h if necessary.

So appealing again to Theorem 1, and also using Theorem 12, we may suppose

$$a_{11} = ah^2, \quad \text{g.c.d. } (h, 2ad) = 1, \quad a_{1j} = 0 \text{ for } j > 2. \quad (151)$$

We take f into a form which is disjoint and has divisor a by the transformation

$$x_1 = y_1 - a_{12}a^2y_2, \quad x_2 = 2h^2a^3y_2, \quad x_i = a^3y_i \quad (i=3,...,n). \quad (152)$$

It is clear that the equation $f(\mathbf{x}) = a$ which we have to solve goes into one of the shape

$$h^2y_1^2 + ag(y_2,...,y_n) = 1, \tag{153}$$

with g an integral form. Now (153) makes y_1 prime to a, whence (152) makes x_1 prime to a, which with $f(\mathbf{x}) = a$ makes $\mathbf{x}$ primitive.

It suffices therefore to prove that (153) is soluble. To do this we need three properties of the form g (or ag). The first is that ag is either indefinite or negative (if not, the left member of (153) would be definite, as would f). The second is that g is of rank $\geqslant 3$, since $n = n(f) \geqslant 4$; thus Theorem 52 is applicable to ag. The third is that if $p \mid h$, implying $p \nmid 2ad$, then ag represents an integer congruent to 1 modulo p. To prove this third property it suffices (using the second) to show that $p \nmid d(ag_1)$, where

$$g_1 = g(0, y_3, ..., y_n).$$

For Theorems 29 and 30 show that $f \equiv 1 \pmod{p}$ is certainly soluble if $n(f) \geqslant 2$ and $p \nmid d(f)$. Now $ag_1 = a^5f_1$, where

$$f_1 = f(0, 0, x_3, ..., x_n);$$

so it suffices to prove $p \nmid d(f_1)$. This follows from (151) and $p \nmid d$, which with $p \mid h$ give $d = d(f) \equiv a_{12}^2d(f_1) \not\equiv 0 \pmod{p}$.

This last property shows that ag represents properly an integer, say b, congruent to 1 modulo h. Now Theorem 51 (ii) shows that $ag \equiv a' \pmod{m}$ is soluble (properly) for every $m \neq 0$ if $a' \equiv b \pmod{m'}$. Here m' is an integer $\neq 0$ whose precise value is immaterial; it can be taken to be $4d(ag)\,\delta$, where δ is the product of the distinct primes dividing $d(ag)$. We choose r, $q(\neq 0)$, with q prime to h so that $m' \mid h^rq$. Now we appeal to Theorem 52, with g, hq for f, m_1. Thus we see that sufficient conditions for ag to represent a' are:

$$a' < 0, \quad a' \equiv b \pmod{h^rq}, \quad a' \equiv 0 \pmod{k^2},$$

where every two of h, q, k are coprime.

Clearly it suffices to choose y_1 so that $1 - h^2 y_1^2 = a'a''^2$, for some $a'' \neq 0$ and a' satisfying the foregoing conditions. To do this it suffices to solve for y_1, c the congruence

$$1 - h^2 y_1^2 \equiv 16bc^2 q^2 k^2 \pmod{16h^r c^2 q^3 k^2},$$

subject to the trivial condition $1 - h^2 y_1^2 < 0$.

We first choose c, with $c \neq 0$, c prime to h, $c \leqslant h^r$, so that this congruence is satisfied modulo h^r by $y_1 \equiv 0$. Then since h is prime to $2cqk$, we may make an obvious change of variable, and we see that we need only solve

$$z_1^2 \equiv 1 - 16bc^2 q^2 k^2 \pmod{16c^2 q^3 k^2}.$$

This congruence breaks up into congruences to prime power moduli p^t, with the right member congruent to 1 $(\bmod\, p)$ always, and congruent to 1 modulo 8. They are therefore all trivially soluble, and the proof of the theorem is complete.

COROLLARY. *If* $|s| = n \geqslant 4$ *and f represents properly an integer congruent to a ($\neq 0$) modulo d, but does not represent a properly, then f represents* ap^2 *properly for every sufficiently large prime p.*

Proof. We seek a solution of (153) with p^2 in place of 1 on the right. From such a solution we derive a solution of $f(\mathbf{x}) = ap^2$ with g.c.d. $(x_1, ..., x_n) = 1$ or p. To find such a solution we have to replace $1 - h^2 y_1^2$ by $p^2 - h^2 y_1^2$ in the congruences. Suppose p so large that it does not divide $2hqk$, and $> h^r$, so that c can be chosen $\leqslant h^r$ and prime to p. Then the congruences can be satisfied with $|y_1| \leqslant 8h^r c^2 q^3 k^2 \leqslant 8h^{3r} q^3 k^2$. Instead of the trivial inequality $1 - h^2 y_1^2 < 0$ we need (since ag is positive) $p^2 - h^2 y_1^2 > 0$. This can be satisfied if p is large enough; indeed, it is (crudely) sufficient (throughout the argument) to take $p > 8h^{3r+1} q^3 k^2$.

CHAPTER 6

RATIONAL TRANSFORMATIONS

1. Equivalence of rational matrices

In this chapter we shall use the symbols $R = (r_{ij}), R_1, \ldots$ to denote non-singular square matrices with rational elements, of rank n (unless otherwise indicated by the context). $T, T_1, \ldots$ will denote square matrices with integral elements and determinant ± 1. It will sometimes be convenient to denote by E_{hk} a square matrix with all elements 0 except a 1 in the (h, k) position.

R, R_1 are said to be EQUIVALENT, in symbols $R \sim R_1$, if there exist T, T_1 such that $R_1 = TRT_1$. Plainly this is a reflexive, symmetric and transitive relation (cf. (16)). To see its relevance to quadratic forms, note that if $f^R = f'$ and $R \sim R_1$, then we have $\phi^{R_1} = \phi'$ for two forms $\phi \, (=f^{T^{-1}}), \phi' \, (=f'^{T_1})$ equivalent to f, f' respectively.

We shall find a canonical shape for a rational R under equivalence. We begin by showing that every R is equivalent to one with $r_{11} > 0, r_{11} \,|\, r_{ij}$ for all i, j. To prove this, consider the rational r'_{11} with the property that some given R is equivalent to an R_1 with r'_{11} as its $(1, 1)$ element. Since we assume $|R| \neq 0$, some $|r_{ij}|$ is positive. Plainly every $|r_{ij}|$ is a value of r'_{11}, so we may suppose $r_{11} > 0$. There is a least positive r'_{11}, since all the r'_{11} are plainly multiples of some non-zero rational. So we may assume (using the transitive property) that r_{11} is the least positive r'_{11}. Multiplying on the left by a suitable T, we may clearly assume that $0 \leqslant r_{i1} < r_{11}$ for $i = 2, \ldots, n$; now the assumption that r_{11} is the least positive r'_{11} gives $r_{i1} = 0$ for $i = 2, \ldots, n$. Other values of r'_{11} are (for $i, j = 2, \ldots, n$ and for any integer h)

$$r_{1j} + hr_{11} \quad \text{and} \quad r_{1j} + r_{ij} + hr_{11}.$$

The assumption that these are all $\leqslant 0$ or $\geqslant r_{11}$ shows that R has the desired property.

It now follows that (with a slight change of notation) every R satisfies

$$R \sim D(R) = [r_1, \ldots, r_n], \tag{154}$$

since we can remove the r_{1j} $(j \geqslant 2)$ by obvious operations. Here the rational numbers $r_i = r_i(R)$ satisfy

$$
\left.
\begin{aligned}
r_i &> 0 \quad (i = 1, \ldots, n), \\
r_i &\mid r_{i+1} \quad (i = 1, \ldots, n-1).
\end{aligned}
\right\} \tag{155}
$$

We show that $D(R)$ depends (as the notation suggests) only on R, or in other words that the r_i are invariants of R under equivalence. Consider first r_1; it is plainly the greatest positive rational dividing all the elements of $D(R)$. It has the same property in relation to R, since the elements of R, $D(R)$ are linear combinations, with integral coefficients, of those of $D(R)$, R; its invariance follows. Next consider, instead of the elements of R, $D(R)$, the determinants of their $i \times i$ sub-matrices $(i = 2, \ldots, n)$; the greatest positive rational of which these are all integral multiples is $r_1 \ldots r_i$, and so $r_1 \ldots r_i$ is invariant.

The matrix $D(R)$ factorizes commutatively:

$$
D(R) = \prod_p D_p(R), \tag{156}
$$

where the product may be supposed taken over $p = 2, 3, \ldots$ but only finitely many factors differ from the identity. Here we have, by (155),

$$
D_p(R) = [p^{\theta_1}, \ldots, p^{\theta_n}], \tag{157}
$$

for integers $\theta_i = \theta_i(p, R)$ satisfying

$$
\theta_1 \leqslant \ldots \leqslant \theta_n. \tag{158}
$$

It is clear that $D_p(R)$ is invariant not only under equivalence but also under multiplication, on either side, by any R_1 with qR_1 integral and $p \nmid |qR_1|$, for any $q \not\equiv 0 \pmod{p}$. It follows that

$$
D(RR_1) = D(R) D(R_1), \quad \text{if g.c.d. } (\text{den } R^{\pm 1}, \text{den } R_1^{\mp 1}) = 1, \tag{159}
$$

for all four choices of the ambiguous signs. Here den R denotes the DENOMINATOR of R, defined in the obvious way as the least positive integer whose product with R is an integral matrix. To deduce (159) from the preceding remark, notice that

$$
(\text{den } R) R(\text{den } R^{-1}) R^{-1} = (\text{den } R) (\text{den } R^{-1}) I,
$$

whence

$$
|(\text{den } R) R| \text{ divides } (\text{den } R)^n (\text{den } R^{-1})^n.
$$

The condition $(159)_2$ can be simplified by omitting the ambiguous signs if we assume (as we usually shall) that $|R| = \pm 1$, $|R_1| = \pm 1$. For then

$$R^{-1} = |R| \operatorname{adj} R \quad \text{and so} \quad \operatorname{den} R^{-1} | \operatorname{den} \operatorname{adj} R | (\operatorname{den} R)^{n-1}.$$

I define the WEIGHT, $w(R)$, of R to be the least positive integer such that $w(R) |R_1|$ is integral whenever R_1 is a square sub-matrix of R, of any rank from 1 to n inclusive. It is clear that $w(R)$ is invariant under equivalence and that

$$\operatorname{den} R | w(R) | (\operatorname{den} R)^n. \tag{160}$$

In terms of $D(R)$, den R is the denominator of r_1, and $w(R)$ is the least common denominator of all the $r_1 \dots r_i$ $(i = 1, \dots, n)$.

Finally, I denote by norm R (the NORM of R) the (unique, positive) square-free integer such that (norm R) $w(R)$ is a square. Notice that $w(T) = \operatorname{norm} T = 1$ for every T.

The last remark shows that f cannot be equivalent to f' if there is a prime p such that $f^R = f'$ (with den R prime to $d(f)$) implies that norm R is not a p-adic square. We shall thus later obtain necessary conditions for equivalence which are easy to work with but not implied by any we have found so far.

2. Rational relations between semi-equivalent forms (further properties)

We now obtain an interesting property of the rational transformations proved to exist in Theorem 50:

THEOREM 54. *Suppose that* $|R| = \pm 1$ *and that there exists an integral quadratic form f such that f^R is integral and f has discriminant d prime to $w(R)$. Then $R \sim R^{-1}$. That is, the rational numbers r_i defined by* (154) *and* (155) *satisfy*

$$r_i r_{n+1-i} = 1 \quad for \quad i = 1, \dots, n, \tag{161}$$

whence r_i is a positive integer for $i > \frac{1}{2}n$, and $r_{\frac{1}{2}(n+1)} = 1$ in case n is odd.

Proof. Since $|R| = \pm 1$, the numbers $\theta_i = \theta_i(p, R)$ of (157) and (158) must satisfy

$$\theta_1 + \dots + \theta_n = 0, \tag{162}$$

and (161) follows if we can prove that for every p they also satisfy

$$\theta_i + \theta_{n+1-i} = 0 \quad \text{for} \quad i = 1, \dots, [\tfrac{1}{2}n]. \tag{163}$$

This is vacuous for $n = 1$ and follows from (162) for $n = 2$; thus the theorem becomes significant for $n \geqslant 3$. It may be noted that (161) can be written as $(JD(R))^2 = I$, where the matrix J is the identity with its rows (or columns) written in reverse order.

It may also be remarked that (161) implies for all n

$$\operatorname{den} R = \operatorname{den} R^{-1}, \quad w(R) = w(R^{-1}), \tag{164}$$

and the converse implication holds for $n = 3, 4$.

Supposing the theorem false we choose a prime p for which (163) fails. Let h ($\leqslant \tfrac{1}{2}n$) be the least value of i for which (163) fails; and suppose that

$$\theta_h + \theta_{n+1-h} < 0. \tag{165}$$

To justify this assumption, note that the hypotheses of the theorem are unaffected by putting R^{-1} for R and interchanging f, f^R; this changes the signs of the θ_i and reverses their order. Further, note that we can replace $R = TD(R)T_1$ by $D(R)$ if we replace f, f^R by the equivalent forms related to them by T^{-1}, T_1; so we may suppose $R = D(R)$.

Thus the hypothesis that f^R is integral gives us that $r_i r_j a_{ij}$ is integral for all i, j, whence

$$p^{-\theta_i - \theta_j} \mid a_{ij} \quad \text{for} \quad i = 1, \dots, n, j = 1, \dots, n. \tag{166}$$

Now (165) gives $p \mid w(R)$, so the theorem is proved if we obtain a contradiction with the hypothesis g.c.d. $(d, w(R)) = 1$ by deducing from (165) and (166) that $p \mid d$. We shall indeed deduce $p^2 \mid d$, by constructing a matrix P with $|P| = p^{-1}$, such that f^P, with discriminant $p^{-2}d$, is an integral form.

We take $P = [p^{-1}, \dots, p^{-1}, 1, \dots, 1, p, \dots, p]$, with h elements p^{-1} and $h - 1$ elements p. To prove f^P integral we need

$$\left. \begin{array}{ll} p^2 \mid a_{ij} & \text{for} \quad i, j \leqslant h, \\ p \mid a_{ij} & \text{for} \quad i \leqslant h, h < j \leqslant n+1-h, \end{array} \right\} \tag{167}$$

as the coefficients of f^P that are not obviously integral are just $p^{-2}a_{ij}$, $p^{-1}a_{ij}$ for i, j satisfying these conditions.

We now complete the proof of the theorem by noting that (167) follows from (158), (165) and (166).

COROLLARY. *With the hypotheses of Theorem* 54, *the matrix R can be factorized into a product $R_1 R_2$ in such a way that f^{R_1} is an integral form, $w(R) = w(R_1) w(R_2)$, and $w(R_1)$ is any prime dividing $w(R)$.*

Remark. The corollary is vacuous unless $w(R)$ is composite; and the factorization may be carried further if $w(R)$ has more than two prime factors (counting multiplicity).

Proof. Choose any p with $p \mid w(R)$; that is, with

$$\theta_1 = \theta_1(p, R) < 0.$$

Let h, with $1 \leqslant h \leqslant \frac{1}{2}n$, be the number of i (1 included) for which $\theta_i = \theta_1$; whence h is also the number of i with $\theta_i = \theta_n$, by (163). As in the proof of the theorem we assume $R = D(R)$ and deduce (166).

We write $R = D(R) = D_p D_1$, where $D_p = D_p(R)$ and D_1 is the product of the other factors in (156). We take

$$R_1 = [I, T][p^{-1}, 1, ..., 1, p],$$
$$R_2 = [p, 1, ..., 1, p^{-1}][I, T^{-1}] D_p D_1,$$

with a suitable T of rank h. Whatever our choice of T, $[I, T^{-1}]$ commutes with D_p by the choice of h, which makes

$$D_p = [p^{\theta_1} I_h, D_0, p^{\theta_n} I_h]$$

for some D_0. Thus $R_2 = R_3 R_4$, with $w(R_4) = w([I, T^{-1}] D_1)$ prime to p and $w(R_3) = w([p, 1, ..., 1, p^{-1}] D_p) = p^{-1} w(D_p)$, using (163) and $\theta_1 < 0$. $w(R_2) = p^{-1} w(R)$ follows easily on using (159).

We have to show that, for suitable T, $f(R_1 \mathbf{x})$ is an integral form. Using (166) we see that $f(R_1 \mathbf{x}) = f^{[I, T]}(p^{-1} x_1, x_2, ..., x_{n-1}, p x_n)$ is integral as far as the terms not involving an x_i with $i > n - h$ are concerned, these being the same as the corresponding ones of $f(p^{-1} x_1, x_2, ..., x_{n-1}, p x_n)$. The only other terms that need consideration are

$$p^{-1} x_1 (a_{1, n+1-h}, ..., a_{1n}) T\{x_{n+1-h}, ..., x_{n-1}, p x_n\}.$$

These can be dealt with by choosing T so that $(a_{1, n+1-h}, ..., a_{1n}) T$ has its first $h - 1$ elements zero. To see that this is possible, use Theorem 1, transposing and putting h for n.

3. Successive rational transformations

The main object of this section is to find a relation between the weights of the rational matrices R_1, R_2, $R_1 R_2$ when R_1 and $R_1 R_2$ both satisfy the conditions of Theorem 54, with the same f. When $w(R_1)$, $= w(R_1^{-1})$ by Theorem 54, is prime to $w(R_2) = w(R_2^{-1})$, it is clear from (159) that $w(R_1 R_2) = w(R_1) w(R_2)$. The case g.c.d. $(w(R_1), w(R_2)) > 1$ is more difficult, and it is convenient to begin with a lemma. It simplifies the statement and proof of the lemma to introduce a temporary notation.

We shall write $R_1, R_2 \mathscr{R} R_3, R_4$ if there exist T_1, T_2, T_3 such that $R_3 = T_1 R_1 T_2^{-1}, R_4 = T_2 R_2 T_3$; the matrices are all supposed square, of the same rank n. Clearly this is a reflexive, symmetric and transitive relation between pairs of rational $n \times n$ matrices. To see its relevance to the problem explained above, suppose f, f', f'' are three forms with $f' = f^{R_1}$, $f'' = f'^{R_2}$. Then there exist three forms ϕ, ϕ', ϕ'', equivalent to f, f', f'' respectively, such that $\phi' = \phi^{R_3}$, $\phi'' = \phi'^{R_4}$.

Some obvious properties of this relation $\mathscr{R}$ are:

(i) $R_1, R_2 \mathscr{R} R_3, R_4$ implies $R_1 \sim R_3$, $R_2 \sim R_4$, $R_1 R_2 \sim R_3 R_4$;

(ii) $R_1 \sim R_3$ implies $R_1, R_2 \mathscr{R} R_3, R_4$ for any R_2 and some $R_4 \sim R_2$;

(iii) $R_1, R_2 \mathscr{R} R_1, T_2 R_2 \mathscr{R} R_1 T_2, R_2$ for any T_2 such that $R_1 T_2 R_1^{-1}$ $(= T_1)$ is integral;

(iv) $R_1, T R_2 \mathscr{R} R_1, R_2$ if $R_2^{-1} T R_2$ is integral.

LEMMA. *Suppose that* R_1, R_2 *are each* $n \times n$, *with determinants* ± 1; *also that* $w(R_2)$ *is a prime* p, *and* $w(R_1)$ *a power of* p. *Then there exist a diagonal matrix* D *and an integer* q, *such that the diagonal elements of* D *are all powers of* p *and*

$$R_1, R_2 \mathscr{R} D, T_0 P \mathscr{R} D T_0, P,$$

where $\quad T_0 = \left[\begin{pmatrix} 1 & 0 \\ pq & 1 \end{pmatrix}, 1, ..., 1 \right], \quad P = [p^{-1}, p, 1, ..., 1].$

Proof. We remark first that $w(R_2) \neq 1$ implies $R_2 \neq \pm I, n \geqslant 2$; if $n = 2$, P is $[p^{-1}, p]$, and similarly for T_0. From $w(R_2) = p$ and $|R_2| = \pm 1$ it is clear that $R_2 \sim P$, so by property (ii) above we may suppose

$$R_1, R_2 \mathscr{R} D, TP; \quad D = [p^{\theta_1}, ..., p^{\theta_n}]. \tag{168}$$

Here the sum of the θ_i must be 0, but we do not assume that they are in ascending order.

At least one of the elements t_{i1} of the first column of T must be prime to p. Of all the $t_{i1} \not\equiv 0 \pmod{p}$ choose one for which θ_i is least, say t_{h1}. We may assume $h = 1$ by permuting the rows of T and the columns of D in the same way, and then permuting the rows of D so that it remains diagonal. Now we can suppose that $p \mid t_{i1}$ for $i > 1$, $p \nmid t_{11}$. For we can add ct_{11} to t_{i1}, c a suitably chosen integer, by multiplying T on the left by $I + cE_{i1}$ (see §1 for definition of E_{hk}). This operation is needed only for $\theta_i \geqslant \theta_1$, by construction, and then is permissible by property (iii) above and (168), since $D(I + cE_{i1}) D^{-1} = I + cp^{\theta_i - \theta_1} E_{i1}$ is integral.

Now by a similar argument there is a $k \neq 1$, which we may suppose $= 2$, such that the kth row of the matrix obtained from T by omitting the first two columns is congruent modulo p to a linear combination of rows for which $\theta_i \geqslant \theta_k = \theta_2$, and $i \neq 1$. (The rows of an $(n-1) \times (n-2)$ matrix are linearly dependent.) We may suppose that this row is divisible by p.

These two steps give us that T in (168) may be supposed, denoting its general element by t_{ij}, to satisfy $p \nmid t_{11}$, $p \mid t_{i1}$ if $i \neq 1$, $p \mid t_{2j}$ if $j > 2$. There is therefore an integer q with $pqt_{11} \equiv t_{21} \pmod{p^2}$. Multiplying T by T_0^{-1} on the left, and D by T_0 on the right, we may replace (168) by

$$R_1, \; R_2 \mathscr{R} D T_0, \; TP, \tag{169}$$

with
$$p^2 \mid t_{21}, \quad p \mid t_{i1} \;\; (i \neq 1), \quad p \mid t_{2j} \;\; (j \neq 1). \tag{170}$$

The lemma follows from (169) and property (iv) above, on noting that (170) makes $P^{-1}TP$ integral.

THEOREM 55. *Suppose that* $|R_1|^2 = \ldots = |R_\rho|^2 = 1$, $\rho \geqslant 2$, $w(R_1) \ldots w(R_\rho)$ *is prime to* $d \neq 0$, *and* $R_1, \ldots, R_\rho$ *all have the same rank* n. *Suppose also that there is an integral, n-ary form f such that* $f(R_1 \ldots R_i \mathbf{x})$ *is integral for* $i = 1, \ldots, \rho$. *Then*

$$w(R_1) \ldots w(R_\rho) \, w(R_1 \ldots R_\rho) \text{ is a perfect square.}$$

Proof. It is clearly sufficient to consider the case $\rho = 2$. The case $w(R_2) = 1$ is trivial. If $w(R_2)$ is composite, we may by the corollary to Theorem 54 write $R_2 = R_3 R_4$, $w(R_3)$ prime and

$w(R_2) = w(R_3) w(R_4)$, with $f(R_1 R_3 \mathbf{x})$ integral. The result follows for R_1, $R_3 R_4$ if proved for R_1, R_3 and for $R_1 R_3$, R_4. Repeating this argument, we see that we may suppose $w(R_2)$ to be a prime p (implying $n \geqslant 2$).

Now we use the corollary to Theorem 54 to factorize R_1. A similar argument shows that it suffices to consider two cases: (i) $w(R_1)$ prime to $w(R_2)$, (ii) $w(R_1) w(R_2)$ a prime power. The first case gives $w(R_1) w(R_2) w(R_1 R_2) = w(R_1 R_2)$ as noted at the beginning of this section. Finally, therefore, we may suppose $w(R_1)$ a power of $p = w(R_2)$, and use the lemma.

There is a form, say g, with coefficients b_{ij}, with $p \nmid d = d(g)$, such that g is integral and remains so on transformation by either D^{-1} or $T_0 P$. We write $g_2 = g(x_1, x_2, 0, ..., 0)$ ($= g$ if $n = 2$), and $g_{n-2} = g(0, 0, x_3, ..., x_n)$ if $n > 2$.

From the hypothesis that $g(T_0 P \mathbf{x})$ is integral it follows easily, using

$$g(T_0 P \mathbf{x}) = g(p^{-1} x_1, q x_1 + p x_2, x_3, ..., x_n),$$

that $p \mid b_{1j}$ for $j \neq 2$. This gives

$$d \equiv b_{12}^2 d(g_{n-2}) \equiv d(g_2) d(g_{n-2}) \pmod{p},$$

whence $\qquad\qquad p \nmid d(g_2), \quad p \nmid d(g_{n-2}).$

Now when g is transformed by D^{-1}, g_2, g_{n-2} are transformed, into integral forms, by matrices with determinants

$$p^{\pm \theta}, \quad \theta = -\theta_1 - \theta_2 = \theta_3 + ... + \theta_n.$$

This is possible, since the transforms would have discriminants $p^{2\theta} d(g_2)$, $p^{-2\theta} d(g_{n-2})$, and $p \nmid d(g_2) d(g_{n-2})$, only if $\theta = 0$, or $\theta_1 = -\theta_2$.

In case $q = 0$, this gives the result. For write

$$w([p^{\theta_3}, \overset{\bullet}{...}, p^{\theta_n}]) = w(= 1 \text{ if } n = 2),$$

and $|\theta_1| = \alpha$. We see that

$$w(R_1) = p^\alpha w, \quad w(R_2) = p, \quad w(R_1 R_2) = w(DP) = p^{\alpha \pm 1} w.$$

We may therefore suppose $q \neq 0$. The argument of the lemma shows however that this case can be avoided unless $D T_0 D^{-1}$ is non-integral, which requires $\theta_1 > \theta_2 = -\theta_1$, $\theta_1 > 0$. The hypothesis that the form g^{D-1} is integral thus gives $p^2 | b_{11}$. A little calculation now shows that with $g^{T_0 P}$ integral, implying

$$g_2(p^{-1}x_1, qx_1 + px_2)$$

integral, we must have $p | b_{12}q$, $p | b_{12}$, $p | d(g_2)$. This contradicts $p \nmid d(g_2)$ previously proved.

It follows at once, by the definition of the norm, that with the hypotheses of the theorem

$$(\text{norm } R_1...R_\rho)(\text{norm } R_1)...(\text{norm } R_\rho) \text{ is a square.} \quad (171)$$

This formula determines norm $R_1...R_\rho$ uniquely when norm $R_1, ..., \text{norm } R_\rho$ are given, since the norms are square-free by definition. In applying the theorem, we shall often take some or all of the R_i to be automorphs of f; and then the requirement that $f(R_1...R_i \mathbf{x})$ be integral becomes trivial.

Note also that the hypothesis $w(R_1)...w(R_\rho)$ prime to d is superfluous in dealing with a prime p not dividing d; thus we have the

COROLLARY. *With the hypotheses of Theorem 55, but omitting the condition that $w(R_1)...w(R_\rho)$ be prime to d, we can conclude that $w_0 w(R_1)...w(R_\rho) w(R_1...R_\rho)$ is a square for some w_0 dividing d.*

4. Reflexions

We repeat here, for convenience, the definition given in (25) of a reflexion $U = U(\mathbf{t})$ of a form f with matrix A:

$$U(\mathbf{t}) = U(\mathbf{t}, f) = U(\mathbf{t}, A) = I - \mathbf{t}\mathbf{t}'A/f(\mathbf{t}), \quad f(\mathbf{t}) \neq 0.$$

Now we observe that when f, with an automorph S, is transformed rationally into f^R, then $R^{-1}SR$ is an automorph of f^R, since

$$f^{RR^{-1}SR} = f^{SR} = f^R \text{ if } f^S = f;$$

that is, S transforms into $R^{-1}SR$. The next formula, which follows at once from (25), shows that reflexions transform into reflexions:

$$U(R^{-1}\mathbf{t}, R'AR) = R^{-1}U(\mathbf{t}, A)R. \quad (172)$$

We can always (see Theorem 3) transform f rationally so as to take $\mathbf{t}$ with $f(\mathbf{t}) \neq 0$ into $\{1, \mathbf{0}\}$, and then, without altering the vector $\{1, \mathbf{0}\}$, split off $a_{11}x_1^2$ rationally provided $a_{11} \neq 0$ (see (20)).

Hence using (172) we can prove some properties of reflexions by considering the case $f = a_{11}x_1^2 + g(x_2, ..., x_n)$, $\mathbf{t} = \{1, \mathbf{0}\}$, in which case $U(\mathbf{t})\mathbf{x} = \{-x_1, x_2, ..., x_n\}$, as is easily seen from (25). In particular, this argument gives us all the properties

$$U(\mathbf{t})\mathbf{t} = -\mathbf{t}, \quad U(\mathbf{t})\mathbf{x} = \mathbf{x} \quad \text{if} \quad \mathbf{t}'A\mathbf{x} = 0, \atop |U(\mathbf{t})| = -1, \quad U^2(\mathbf{t}) = I. \right\} \tag{173}$$

For these properties (for the first two of which see (26)) are easily seen to be rationally invariant.

Now we apply to a reflexion U the theory of §1 of this chapter:

THEOREM 56. *The weight and denominator of a reflexion U are always equal; that is, we have*

$$U \sim [w^{-1}(U), 1, ..., 1, w(U)], = I \quad \text{if} \quad n = 1.$$

If $U = U(\mathbf{t}, f) = U(\mathbf{t}, A)$, with $\mathbf{t}$ integral and primitive, then $w(U) = a^{-1}f(\mathbf{t})$, for some integer a with $a \mid d = d(f)$, af indefinite or positive, and such that $a \mid \mathbf{t}'A$.

Proof. We use (172), with R, R^{-1} integral. We may suppose (Theorem 12) that $\mathbf{t} = \{1, \mathbf{0}\}$ and $a_{11} = f(\mathbf{t}) \neq 0$, $a_{1j} = 0$ if $j > 2$ (the case $n = 1$ is trivial).

Now let $|a| = $ g.c.d.(a_{11}, a_{12}), sgn $a = $ sgn a_{11}, whence af is indefinite or positive since it represents $aa_{11} > 0$, and

$$a \mid \mathbf{t}'A = (2a_{11}, a_{12}, \mathbf{0}').$$

Clearly also d, being a linear combination of a_{11}, a_{12}, is a multiple of a.

The assertions regarding U, $w(U)$ $(= a^{-1}a_{11})$ are now clear on cancelling a in

$$U\mathbf{x} = \{-x_1 + a_{11}^{-1}a_{12}x_2, x_2, ..., x_n\}.$$

Now suppose f, f^R are two integral forms; then also $f^R = f^{UR}$ for any reflexion U of f. We consider whether left multiplication by U will cancel unwanted factors from $w(R)$. We begin (see corollary to Theorem 54) with the case in which R has the property $w(R) = \operatorname{den} R$ that U has just been proved to have.

THEOREM 57. *Suppose that f, f^R are integral forms, of rank $n \geqslant 2$, and that $R \sim [w^{-1}, 1, ..., 1, w]$, for some positive w prime to*

$d = d(f)$. *Then if* $U = U(\mathbf{t}, f)$, $\mathbf{t}$ *integral and primitive, the denominator of* UR *is prime to* w *if* $w^{-1}f(\mathbf{t})$ *is an integer prime to* w, *and* $W^{-1}\mathbf{t}$ *is integral, for a suitably chosen* W *depending only on* f, R *and satisfying* $W \sim [1, w, ..., w]$.

If so, then we have

$$\text{den } UR = w(UR) = w^{-1}\text{den } U.$$

Proof. It is not difficult to see that we may replace f, f^R by equivalent forms without affecting the hypotheses or conclusion. Replacing f, f^R by f^T, f^{RT_1}, we have to replace R, $\mathbf{t}$, W by $T^{-1}RT_1$, $T^{-1}\mathbf{t}T^{-1}W$. We may therefore suppose $R = [w^{-1}, 1, ..., 1, w]$; and with this we shall show that $W = [1, w, ..., w]$ does what is wanted.

With this R, the hypothesis that f^R is integral takes the shape

$$w^2 | a_{11}, w | a_{1j} \quad \text{for} \quad j = 2, ..., n-1. \tag{174}$$

With the chosen W, the requirement that $W^{-1}\mathbf{t}$, $= \mathbf{z}$, say, be integral reduces to $[w, 1, ..., 1]\mathbf{t} \equiv \mathbf{0} \pmod{w}$, or

$$w | t_i \quad \text{for} \quad i = 2, ..., n. \tag{175}$$

We use (174) and (175) to prove

$$wf(\mathbf{t}) \, UR = (f(\mathbf{t})I - \mathbf{t}\mathbf{t}'A)[1, w, ..., w, w^2] \equiv 0 \pmod{w^2}. \tag{176}$$

(176) is clear as far as the last column is concerned. To verify it for the jth column $(1 < j < n)$, we need only to see that w divides the jth column of $\mathbf{t}'A \equiv (t_1, \mathbf{0}')A \pmod{w}$. This is clear from (174) and (175). Considering now the elements, other than the first, of the first column, we need only, because of (175),

$$w | \mathbf{t}'A[1, 0, ..., 0] \equiv (1, \mathbf{0}')A[1, 0, ..., 0] \pmod{w};$$

(174) gives us this. The leading element of the matrix in (176) is

$$f(\mathbf{t}) - t_1\mathbf{t}'\{2a_{11}, a_{21}, ..., a_{n1}\}.$$

By (174) and (175) this is congruent modulo w^2 to

$$a_{n1}t_1t_n - t_1t_na_{n1} \equiv 0 \pmod{w^2};$$

hence (176) is proved.

The first assertion now follows as (176) gives den $UR | w^{-1}f(\mathbf{t})$, prime to w by hypothesis. $w(UR)$ is also prime to w by (160). The second assertion is now straightforward. Clearly

$$w(UR)\,|\,w(U)\,w(R) = w\,\mathrm{den}\,U,$$

and $\qquad w(U) = w(URR^{-1})\,|\,w(UR)\,w(R^{-1}) = w(UR)\,w,$

and so with $w(UR)$ prime to w we have $w(UR) = w^{-1}\,\mathrm{den}\,U$; and a similar argument applies to the denominator.

The last two theorems show that to construct useful reflexions we have to find vectors $\mathbf{t}$ which (i) give convenient values to $f(\mathbf{t})$ and (ii) satisfy certain congruence conditions. One of these conditions is $\mathbf{t} = W\mathbf{z}$, with the W of Theorem 57 and integral $\mathbf{z}$. This can be got rid of by considering instead of f the form f^W; or $w^{-1}f^W$, which is plainly integral. The other is the condition $a\,|\,\mathbf{t}'A$ of Theorem 56. This may be got rid of in a similar way, or by working (for primitive f) with $a = \pm 1$. There remains the problem of finding a vector, say $\mathbf{x}$, such that $\phi(\mathbf{x})$ has some desired value, or at least a value satisfying certain conditions, ϕ being a form of the same rank n as f. This is unnecessary in the trivial case $n = 1$, and difficult for $n = 2, 3$. It becomes easier for $n \geqslant 4$, especially if f is indefinite. For $n = 3$, we show in the next section that—considering the general automorph instead of restricting ourselves to reflexions—we can get what we want by a similar construction with a *quaternary* ϕ.

5. Hermite's formula for the automorphs of a ternary form

We can set up a 1-1 correspondence between all skew 3×3 and all 3×1 matrices by writing

$$Z(\mathbf{x}) = \begin{pmatrix} 0 & -x_3 & x_2 \\ x_3 & 0 & -x_1 \\ -x_2 & x_1 & 0 \end{pmatrix} \quad \text{for} \quad \mathbf{x} = \{x_1, x_2, x_3\}. \quad (177)$$

Clearly $\qquad \mathbf{x}'Z(\mathbf{x}) = (0, 0, 0), \quad Z(\mathbf{x})\,\mathbf{x} = \{0, 0, 0\}. \quad (178)$

It is not difficult to verify the identity

$$Z(\mathbf{y})\,(\mathrm{adj}\,A)\,Z(\mathbf{x}) = A\mathbf{x}\mathbf{y}'A - (\mathbf{x}'A\mathbf{y})\,A; \quad (179)$$

from obvious considerations of linearity and symmetry it suffices to take $\mathbf{x} = \{1, 0, 0\}$ and $\mathbf{y} = \mathbf{x}, \{0, 1, 0\}$.

Using this notation we state and prove

THEOREM 58. *Let f be a ternary form, $\mathbf{x} = \{x_1, x_2, x_3\}$ integral, and x_4 any integer such that $x_4^2 \neq df(\mathbf{x})$, where $d = d(f) = -\frac{1}{2}|A|$, $d \neq 0$, $A = A(f)$.*

Define S by

$$(x_4^2 - df(\mathbf{x})) S = (x_4^2 + df(\mathbf{x})) I + x_4(\operatorname{adj} A) Z(\mathbf{x}) - d\mathbf{x}\mathbf{x}'A,$$
(180)

with $Z(\mathbf{x})$ as defined in (177).

Then S is a rational automorph of f, and $|S| = 1$.

Proof. We have to show that $S'AS = A$. On multiplying by $(x_4^2 - df(\mathbf{x}))^2$ this becomes an algebraic identity, and is easily verified directly, making use of (178), the case $\mathbf{y} = \mathbf{x}$ of (179), and $\mathbf{x}'A\mathbf{x} = 2f(\mathbf{x})$, $|A| = -2d$.

Now we have to prove $|S| = 1$; and this amounts to verifying an algebraic identity in the x_i. Suppose first that $f(\mathbf{x}) \neq 0$. Then if $x_4 = 0$ we have $S = -U(\mathbf{x})$, $|S| = -|U(\mathbf{x})| = 1$, by $(173)_3$. Next suppose x_4 divisible by a prime p with $p \nmid 2df(\mathbf{x})$. Clearly

$$S \equiv -U(\mathbf{x}), \quad |S| \equiv -|U(\mathbf{x})| = 1 \pmod{p}. \quad |S'AS| = |A| \neq 0$$

gives $|S|^2 = 1$, so $|S| = 1$. Now we may remove first the restriction $p \mid x_4$ and then the restriction $f(\mathbf{x}) \neq 0$, since it suffices to verify a polynomial identity for a finite number of values of the variable. (In using this elementary result we keep three of the x_i fixed at each step.) The proof is now complete and we deduce

THEOREM 59. *Let f, w, R, W be as in Theorem 57, and $n = 3$. Let $t_1, \ldots, t_4$ be integers such that*

$$\text{g.c.d.}\ (t_1, \ldots, t_4) = \text{g.c.d.}\ (t_1, t_2, t_3, w) = 1;$$
(181)

and writing $\mathbf{t} = \{t_1, t_2, t_3\}$ suppose that

$$w \mid f(\mathbf{t}), \quad \text{g.c.d.}\ (w^{-1}f(\mathbf{t}) - dw^3t_4^2, 2dw) = 1.$$
(182)

Define V by

$$(f(\mathbf{t}) - dw^4t_4^2)\,V = (f(\mathbf{t}) + dw^4t_4^2)\,I + w^2t_4(\operatorname{adj} A)\,Z(\mathbf{t}) - \mathbf{t}\mathbf{t}'A.$$
(183)

Then V is a rational automorph of f, with $|V| = -1$. Moreover,

$$\operatorname{den} VR = w^{-1} \operatorname{den} V, \quad \operatorname{den} V = |f(\mathbf{t}) - dw^4 t_4^2|.$$

Proof. The first two assertions follow from Theorem 58; for with an obvious notation $V = -S(\mathbf{t}, dw^2 t_4)$. From (182) we have $w^{-1}f(\mathbf{t})$ integral, prime to w, so (183) gives $wV \equiv wU(\mathbf{t}) \pmod{w^2}$. Using Theorem 57, we deduce that $w^{-1} \operatorname{den} V$ is integral and prime to w, and that the third assertion is true. (We have only $(181)_2$ in place of the stronger hypothesis of Theorem 57 that $\mathbf{t}$ is primitive, but this clearly does not affect the argument.)

The denominator of V is clearly a divisor of $f(\mathbf{t}) - dw^4 t_4^2$. By what we have already proved, the fourth assertion follows if we can deduce a contradiction from the assumption that some prime p divides the right member of (183) and also divides $f(\mathbf{t}) - dw^4 t_4^2$, but not w. With this assumption, $p \nmid 2dw$, by (182).

A little calculation shows that the trace ($=$ sum of diagonal elements) of the second, third term on the right of (183) is $0, -2f(\mathbf{t})$. The hypotheses regarding p therefore give us

$$p \mid 3f(\mathbf{t}) + 3dw^4 t_4^2 - 2f(\mathbf{t}), \quad p \mid 4dw^4 t_4^2, \quad p \mid t_4, \quad p \mid f(\mathbf{t}).$$

On multiplying (183) on the right by $\operatorname{adj} A$, we have

$$p \mid (f(\mathbf{t}) + dw^4 t_4^2) \operatorname{adj} A + w^2 t_4 (\operatorname{adj} A) Z(\mathbf{t}) (\operatorname{adj} A) + 2d\mathbf{t}\mathbf{t}',$$

since $A(\operatorname{adj} A) = |A| I = -2dI$. Using $p \mid t_4$ and noticing that the second term on the right is skew since $Z(\mathbf{t})$ is so, we may add this congruence to its transpose, which gives, since the other terms are symmetric, $p \mid 2f(\mathbf{t}) \operatorname{adj} A + 4d\mathbf{t}\mathbf{t}'$. With $p \mid f(\mathbf{t})$ and $p \nmid 2d$ this gives $p \mid \mathbf{t}\mathbf{t}'$, $p \mid \mathbf{t}$, which with $p \mid t_4$ contradicts (181). This completes the proof.

6. Construction of automorphs

The object of this section is to construct automorphs (reflexions or products of reflexions except for $n = 3$, when we also use Theorem 59) which can be used to improve Theorem 50. Throughout this section f is an integral form, of rank $n \geqslant 2$ and discriminant $d \neq 0$; and we assume that f is integral and either indefinite or positive. R is a rational $n \times n$ matrix with determinant ± 1, such that $w(R)$ (> 1) is prime to d and f^R is integral. S denotes an automorph of f, with $w(S)$ prime to d. We state the results as lemmas.

LEMMA 1. *Suppose* $R \sim [w^{-1}, 1, ..., 1, w]$ *and let w' be any positive integer, prime to dw, such that* $f(\mathbf{y}) \equiv ww'$ (mod $4d$) *is soluble* (*with integral* $\mathbf{y}$). *In case $n = 2$, impose on w' the further condition that d is a p-adic square for every p dividing w'.*

Then for every integer $m_1 \neq 0$ there is an S with $|S| = -1$ such that, for some h prime to $dww'm_1$,

$$SR \sim [w'^{-1}h^{-2}, 1, ..., 1, w'h^2].$$

Proof. We shall take $S = U(\mathbf{t})$, for some $\mathbf{t}$ with $f(\mathbf{t}) = ww'h^2$, for an h with the desired properties. With such an S, Theorem 56 shows that we have den $S = w(S) = f(\mathbf{t})$. Then Theorem 57 gives the desired result, if we impose on $\mathbf{t}$ the condition of that theorem that $W^{-1}\mathbf{t}$ be integral. We may, however, as in the proof of that theorem, assume that $R = [w^{-1}, 1, ..., 1, w]$, so that this condition reduces to (175), while the hypothesis that f^R is integral takes the shape (174). Thus we have to solve the equation

$$f(x_1, wx_2, ..., wx_n) = ww'h^2,$$

with $\mathbf{x}$ primitive. Necessarily this equation gives x_1 prime to w, so $\mathbf{t} = \{x_1, wx_2, ..., wx_n\}$ will be primitive and give the desired $U(\mathbf{t})$.

Theorem 48 shows that we can solve this equation, with a suitable h, if the congruence

$$f(x_1, wx_2, ..., wx_n) \equiv ww' \ (\mathrm{mod}\, m)$$

is soluble for every $m \neq 0$ (with primitive $\mathbf{x}$). (It suffices indeed to take $m = 4(ww')^n \, dm_1$, but this is not important.) We use here the assumption that f has values with the sign of $ww' > 0$.

The congruence above is trivial for m prime to w. For with m a power of d it can obviously be satisfied by making

$$\{x_1, wx_2, ..., wx_n\} \equiv q\mathbf{y}$$

for some suitable q (clearly g.c.d. $(y_1, ..., y_n)$ is prime to d). And with m a power of a prime p not dividing dw, it is enough to solve $f \equiv ww'$ (mod m), which is clearly possible by Theorems 29 and 30, and the restriction on w' in case $n = 2$.

It suffices therefore to take $m = w^r$, r any positive integer, and, putting $x_2 = \ldots = x_{n-1} = 0$, to solve

$$f(x_1, 0, \ldots, 0, wx_n) \equiv ww' \pmod{w^r}.$$

Now we have (see (174)) $w^2 \mid a_{11}$, $w \mid a_{1j}$ for $j = 2, \ldots, n-1$. So w, being prime to d, must be prime to a_{1n}. Now writing the last congruence as

$$w^{-1}a_{11}x_1^2 + a_{1n}x_1x_n + wa_{nn}x_n^2 \equiv w' \pmod{w^{r-1}},$$

we see easily that it is soluble; the form on the left has a discriminant prime to w and a p-adic square for every $p \mid w$. This completes the proof. It may be remarked that the discriminant just mentioned is that of f if $n = 2$, whence w necessarily satisfies the additional condition imposed on w' in that case.

LEMMA 2. *Suppose $n \geq 3$, w' odd. Then the h of Lemma 1 may be chosen to be either 1 or a prime not dividing $dww'm_1$; the choice $h = 1$ is always possible if f is indefinite.*

Proof. In case $n \geq 4$ we have only to modify the argument of Lemma 1 by using instead of Theorem 48 the stronger Theorem 53, if f is indefinite, or its corollary if not.

In case $n = 3$, we see by Theorem 59 that instead of solving $f(x_1, wx_2, wx_3) = ww'h^2$, it would have sufficed, in the proof of Lemma 1, to solve (with primitive $\mathbf{x}$)

$$f(x_1, wx_2, wx_3) - dw^4x_4^2 = ww'h^2, \quad \text{odd.}$$

Then with $t_1, \ldots, t_4 = x_1, wx_2, wx_3, x_4$ it is clear that (182) holds; and the argument of Lemma 1 shows that x_1 is prime to w, giving (181).

To show that this Diophantine equation is soluble with the present restrictions on h, we again appeal to Theorem 53 (or its corollary), and we have only to show the solubility of

$$f(x_1, wx_2, wx_3) - dw^4x_4^2 \equiv ww' \pmod{4w^8d^2},$$

with g.c.d. $(x_1, \ldots, x_4) = 1$. The modulus here is the discriminant of the form on the left. It is sufficient to solve instead

$$f(x_1, wx_2, wx_3) \equiv ww' \pmod{4w^8d^2}, \quad \text{g.c.d. } (x_1, x_2, x_3) = 1.$$

Now the argument of Lemma 1 shows that this is possible, so the proof is complete.

LEMMA 3. *Suppose $R = R_1 R_2$, with $R_1 \sim R_2$, f^{R_1} also integral, and $w(R_1) = \operatorname{den} R_1$. Then f, f^R are equivalent if f is indefinite and $n \geqslant 3$. If $n = 2$ or f is definite there is an S with $|S| = 1$ and*

$$SR \sim [h^{-2}, 1, ..., 1, h^2],$$

where h may be chosen prime to any assigned non-zero integer in any case, and prime if $n \geqslant 3$.

Proof. We have by hypothesis $R_1 \sim R_2 \sim [w^{-1}, 1, ..., 1, w]$, for some w. Choose w', positive and prime to dw, and such that $f(\mathbf{y}) \equiv ww' \pmod{4d}$ is soluble; some such w' exists since we assume f primitive.

Now apply Lemma 1, with R_1 for R. With a slight change of notation, this gives us an S_1, with $|S_1| = -1$, such that

$$S_1 R_1 \sim (w'^{-1} h_1^{-2}, 1, ..., 1, w' h_1^2),$$

where (taking $m_1 = 1$) we have h_1 prime to dww'. Clearly this gives

$$\begin{aligned} S_1 R = S_1 R_1 R_2 &\sim [w'^{-1} h_1^{-2}, 1, ..., 1, w' h_1^2][w^{-1}, 1, ..., 1, w] \\ &= [(ww' h_1^2)^{-1}, 1, ..., 1, ww' h_1^2], \end{aligned}$$

by (159).

Now the solubility of $f(\mathbf{y}) \equiv ww' \pmod{4d}$ clearly implies that of $f(\mathbf{y}) \equiv ww' h_1^2 \pmod{4d}$. So we may apply Lemma 1 again, with $S_1 R$ in place of R and $ww' h_1^2$, 1 in place of w, w'. This gives us an S_2 with $|S_2| = -1$ and $S_1 S_2 R \sim [h^{-2}, 1, ..., 1, h^2]$, with h prime to any prescribed multiple of dw. For $n = 2$ this gives us (with $S = S_1 S_2$) all that is asserted. For $n \geqslant 3$ we use Lemma 2 at the second step. In the indefinite case $h = 1$ gives

$$SR \sim I, \quad f^R = f^{SR} \sim f.$$

7. The norms of automorphs; improvement of Theorem 50

It will be important to know what values norm S can take when S is an automorph of a given n-ary form f, and $w(S)$ is restricted to be prime to $d(f)$. We use the results of the last section to reduce this problem to one on congruences.

THEOREM 60. *Let f be an integral, n-ary form with rank $n \geqslant 2$ and discriminant $d \neq 0$. Let q be a square-free positive integer congruent to a square modulo d, and prime to d. Suppose also, in case*

$n = 2$, *that d is a p-adic square for every prime p dividing q. Then f has a rational automorph S with denominator prime to d, determinant* 1, *and norm q. Further, the denominator of S may be supposed prime to m_1, for any assigned $m_1 \neq 0$, prime to q.*

Proof. It is easy to see that if the theorem is true for $f = f_0$ it is also true for $f = cf_0$, c any non-zero integer prime to q. (The final restriction on S is put in mainly so that this may be so.) We may therefore suppose f primitive and either indefinite or positive, and appeal to Lemma 1 of the last section.

We first take $R = I$, $w = 1$, and choose a w' such that $f(\mathbf{y}) \equiv w'$ (mod $4d$) is soluble. Thus we can construct a reflexion U_1 with

$$U_1 \sim [w'^{-1}h_1^{-2}, 1, \ldots, 1, w'h_1^2],$$

h_1 prime to $qdwm_1$ (replacing m_1 by qm_1 in Lemma 1).

Now suppose for the moment that the congruence $f(\mathbf{y}) \equiv qw'$ (mod $4d$) is soluble, with integral $\mathbf{y}$, whence so too is $f(\mathbf{y}) \equiv qw'h_1^2$ (mod $4d$). Then we can apply Lemma 1 again, with $R = U_1$, $w'h_1^2$ for w, and q for w'. We clearly obtain the desired result by putting $S = U_2U_1$ on constructing a reflexion U_2 so that

$$U_2U_1 \sim [q^{-1}h^{-2}, 1, \ldots, 1, qh^2],$$

with h prime to dm_1.

It remains only to show that if we choose w' suitably the congruence $f(\mathbf{y}) \equiv qw'$ (mod $4d$) will be soluble (and to remark that in case $n = 2$ we may suppose w' to satisfy the additional condition imposed in that case on q). With the hypothesis that q is congruent to a square modulo d, and the solubility by construction of $f(\mathbf{y}) \equiv w'$ (mod $4d$), it is clearly sufficient to prove the solubility of $f(\mathbf{y}) \equiv qw'$ modulo 4 if $2 \nmid d$, 8 if $2 \mid d$ (in which case qw' is odd), for suitable w'. The case $2 \nmid d$ is trivial, and the other case needs consideration only if $8 \nmid d$. In case $4 \nmid d$, it is easy to see from (117) that f takes all odd residues modulo 8. If $d \equiv 4$ (mod 8) the hypotheses give $q \equiv 1$ (mod 4), and it suffices to find an odd w' so that $f \equiv w'$, $w' + 4$ (mod 8) are both soluble. This again is quite easy, using (117), so the proof is complete.

Now we obtain an improvement on Theorem 50.

THEOREM 61. *Suppose that f, f^R are both integral forms and that $|R| = \pm 1$, den R is prime to $d = d(f)$, and norm $R = 1$, that is,*

$w(R)$ is a square. Then f, f^R are equivalent if f is indefinite and $n = n(f) \geqslant 3$, also in the trivial case $n = 1$. In the remaining cases we have

$$f^R = f^H, \quad H \sim [h^{-2}, 1, ..., 1, h^2],$$

for an h which may be chosen prime to any assigned non-zero integer m in any case, and to be either 1 or a prime not dividing such m in case $n \geqslant 3$.

Proof. By the corollary to Theorem 54 we have $R = R_1...R_{2u}$, with each $w(R_i)$ prime, $w(R_{2i-1}) = w(R_{2i})$ for $i = 1, ..., u$, and all the forms $f(R_1...R_i\mathbf{x})$ integral, $i = 1, ..., 2u$. Clearly $w(R_1)$ prime makes $R_1 \sim [p^{-1}, 1, ..., 1, p] \sim R_2$, where $p = w(R_1) = w(R_2)$; and similarly for $R_3, ..., R_{2u}$.

Apply Lemma 3 of the last section to find an automorph S_1 of f, with $|S_1| = 1$, such that $S_1 R_1 R_2 \sim [h_1^{-2}, 1, ..., 1, h_1^2]$, h_1 prime to dm. Then use the lemma again, this time with $f^{R_1 R_2}$ for f, and find an automorph S_2 of $f^{R_1 R_2}$ with $S_2 R_3 R_4 \sim [h_2^{-2}, 1, ..., 1, h_2^2]$, h_2 prime to dmh_1.

Proceeding in this way we see that we have $f^R = f(H_1...H_u\mathbf{x})$, where for $i = 1, ..., u$ we have $H_i \sim [h_i^{-2}, 1, ..., 1, h_i^2]$, h_i prime to dm, and every two of the h_i are coprime. Using (159) this gives us $H_1...H_u \sim [w^{-2}, 1, ..., 1, w^2]$, w prime to dm. In case $n = 2$ we take $H = H_1...H_u, h = w$.

In case $n \geqslant 3$ we assume without loss of generality that $H_1...H_u = [w^{-2}, 1, ..., 1, w^2]$. The hypothesis that $f^R = f(H_1...H_u\mathbf{x})$ is integral gives us $w^2 \mid a_{1j}$ for $j \neq n$. This implies that f^Q is integral, where $Q = [w^{-1}, 1, ..., 1, w]$, and $Q^2 = H_1...H_u$. Hence one more application of Lemma 3 gives the result.

We shall see that the hypothesis norm $R = 1$ of the last theorem is not very restrictive for large n; but meanwhile we prove

THEOREM 62. *Assume the hypotheses of Theorem 61, except that* norm $R \neq 1$. *Then* $f^R \sim f^{R_0}$, *where*

$$R_0 \sim [(\text{norm } R)^{-1}, 1, ..., 1, \text{norm } R],$$

in case $n \geqslant 3$ *and* f *is indefinite. Otherwise* $f^R \sim f^{HR_0}$, *with* R_0 *as above and* H *as in Theorem 61.*

Proof. Note that norm $R \neq 1$ excludes the trivial case $n = 1$.

Now the corollary to Theorem 54 gives us $R = R_1 R_0$, with $w(R_0) = \text{norm } R$ and $w(R_1)$ a square. We treat the factor R_1 as in Theorem 61 (choosing h prime to norm R). Now norm R being square-free by definition we have

$$R_0 \sim [(\text{norm } R)^{-1}, 1, \ldots, 1, \text{norm } R]$$

as stated. (For a matrix $[r_1, \ldots, r_n]$ satisfying (155) and (161) has non-square-free weight if $r_{n-1} > 1$.)

CHAPTER 7

EQUIVALENCE AND SPINOR-RELATEDNESS

1. The spinor genus

Let f, f' be any two integral forms, f of rank n and discriminant $d\ (\neq 0)$. We shall say that f is SPINOR-RELATED to f', in symbols $f \mathrel{\dot\sim} f'$, if there exists a rational matrix R such that

$$f^R = f', \quad |R| = \pm 1, \quad \text{g.c.d. (den } R, d) = 1, \quad \text{norm } R = 1. \tag{184}$$

Theorem 61 shows that if $f \mathrel{\dot\sim} f'$ then we can, by varying the choice of R, satisfy these conditions with $(184)_3$ replaced, for any given $m \neq 0$, by the stronger

$$\text{g.c.d. (den } R, dm) = 1. \tag{185}$$

Hence we deduce a property of spinor-relatedness which is trivial for other relations considered so far:

$$mf \mathrel{\dot\sim} mf' \quad \text{if and only if} \quad f \mathrel{\dot\sim} f' \quad (m \neq 0). \tag{186}$$

The 'only if' follows immediately from the definition, but the 'if' needs (185), since $d(mf) = m^n d(f)$ may not be prime to den R for R satisfying (184). We note that by (160) and the definition of the norm, the last two of the conditions (184) hold if and only if $w(R)$ is a square and prime to d. We note that

$$d \text{ is a } p\text{-adic square if} \quad n = 2 \quad \text{and} \quad p \mid w(R), \tag{187}$$

by the remark at the end of the proof of Lemma 1 in Chapter 6, §6. Further, if $n = 1$ then (184) implies $R = \pm I, f = f'$; we shall exclude this trivial case and assume $n \geqslant 2$ throughout this chapter.

It is not immediate that spinor-relatedness has the usual properties (16)–(18) of Chapter 1. The reflexive property is clear since norm $I = 1$. This and the permutation property (17) are particular cases of the obvious

$$f \sim f' \quad \text{implies} \quad f \mathrel{\dot\sim} f'. \tag{188}$$

For the symmetric property, note that (184) and Theorem 54 give $R \sim R^{-1}$, norm $R^{-1} =$ norm $R = 1$. For the transitive property, suppose (184) holds as it stands and also with f', f'', R_1 (f'' integral) for f, f', R. Then $(184)_1$ to $(184)_3$ obviously hold with f'', RR_1 for f', R; and since $f' = f^R$ and $f'' = f^{RR_1}$ are both integral Theorem 55 gives us that $w(RR_1)\, w(R)\, w(R_1)$ is a square, whence $w(RR_1)$ is a square. Now (see (9)) take $m = 2d(f_1) \neq 0$ in (185), and note that norm $([R, I]) =$ norm R; we deduce that

$$f \mathrel{\dot\sim} f' \quad \text{implies} \quad f + f_1 \mathrel{\dot\sim} f' + f_1 \qquad (189)$$

for any integral f_1 with non-zero discriminant. This is (18) (for $\mathscr{R} = \mathrel{\dot\sim}$) with a slight change of notation.

The converse of (188) is true in most, though not in all cases. For the case $n \geqslant 3$, $|s| < n$ of Theorem 61 may be restated as

THEOREM 63. *Spinor-relatedness and equivalence coincide for* $n > \max(2, |s|)$; *in other words, two indefinite forms of rank at least 3 are equivalent if and only if they are spinor-related.*

We may say alternatively that the spinor-genus and the class coincide for $n > \max(2, |s|)$, if we define the SPINOR GENUS (as we may since we have proved the reflexive, symmetric and transitive properties) in such a way that $f \mathrel{\dot\sim} f'$ if and only if f, f' are in the same spinor genus. For any n, s, (188) shows that a class is always included in a spinor genus, which therefore consists of one or more classes.

We shall see later that the condition $n > |s|$ in Theorem 63 cannot be omitted. The need for the condition $n > 2$ may be seen by taking

$$f = x_2(9x_1 + x_2), \quad d = 81, \quad f' = x_2(9x_1 + 16x_2),$$
$$R = [\tfrac{1}{4}, 4], \quad w(R) = 4, \quad \text{norm } R = 1.$$

We have $f \mathrel{\dot\sim} f'$ by the definition, but $f \sim f'$ is false since f represents 1 while f' does not.

We now give a necessary and sufficient condition for spinor-relatedness which is more manageable than the definition.

THEOREM 64. *Let f be an integral form with discriminant $d \neq 0$, and R a rational matrix such that f^R is integral, $|R| = \pm 1$, $w(R)$ is* *prime to d; then f, f^R are in the same spinor genus if and only if there exist an integer k and a rational automorph S of f with*

$$k^2 w(R) w(S) \equiv 1 \pmod{d}. \tag{190}$$

Proof. First suppose $f \stackrel{.}{\sim} f^R$. Then there exists Q such that (184) holds with f^R, Q for f', R. Put $S = RQ^{-1}$. $f^Q = f^R$ gives $f^S = f$. Since f, f^S, $= f$, and f^{SQ}, $= f^Q = f^R$, are all integral, and obviously $|S|^2 = |SQ|^2 = 1$ and $w(S)w(Q)$ is prime to d, Theorem 55 gives us that $w(SQ) w(S) w(Q) = w(R) w(S) w(Q)$ is a square. Now $w(Q)$ is a square by our choice of Q; so $w(R) w(S)$ is a square, clearly prime to d, and (190) holds for some k.

Now suppose (190) holds. We can choose a positive, square-free q such that $qw(R) w(S)$ is a square. In case $n = 2$, note that if $p \mid q$ we have $p \mid w(R)$ or $w(S)$. By (187), which clearly holds also with S for R, d is a p-adic square. So by Theorem 60 we can find a rational automorph S_1 with denominator prime to d and with norm q. Since f^{S_1}, $f^{S_1 S}$, $f^{S_1 SR}$, $= f$, f, f^R are all integral, Theorem 55 gives us that $w(S_1 SR) w(S_1) w(S) w(R)$ is a square.

Hence by our choice of S_1 we see that $w(S_1 SR) w(S) w(R) q$ is a square, whence $w(S_1 SR)$ is a square, and norm $S_1 SR = 1$, by our choice of q. It is clear now that (184) holds with $S_1 SR$ in place of R, whence $f \stackrel{.}{\sim} f'$.

It is clear from the definition that

$$f \stackrel{.}{\sim} f' \quad \text{implies} \quad f \simeq f'. \tag{191}$$

We shall see that this implication, like (188), has a true converse in most cases. To show that the converse is not always true, take $f = x_2(5x_1 + x_2)$, $d = 25$, $f' = x_2(5x_1 + 4x_2)$. Then (190), with this f and $R = [\frac{1}{2}, 2]$, reduces to $w(S) \equiv \pm 2 \pmod{5}$; and we shall show that no such S exists. It is enough to show that there is no such S with $|S| = 1$; for any S with $|S| = -1$ can be multiplied by the integral matrix, with determinant -1, of the substitution $x_2 \to -5x_1 - x_2$ without altering its weight. Now it is easy to see that $\mathbf{x} = S\mathbf{y}$ ($|S| = 1$) must take x_2, $5x_1 + x_2$ into $uv^{-1}y_2$, $vu^{-1}(5y_1 + y_2)$ for some coprime integers u, v. The weight, equal to the denominator, of S is now the least common denominator of uv^{-1}, vu^{-1}, $5^{-1}(uv^{-1} - vu^{-1})$, by a simple calculation.

Hence $w(S) = |uv|$ or $5|uv|$. (190) requires $w(S) = |uv| \not\equiv 0$ (mod 5), and this requires $5 \,|\, u^2 - v^2$, whereas $uv \equiv \pm 2$ (mod 5) gives $5 \,|\, u^2 + v^2$.

2. Some possible values for the weight of an automorph modulo d

We prove in this section four theorems which will enable us to simplify Theorem 64.

THEOREM 65. *Every indefinite form f has a rational automorph S with $w(S) \equiv -1$ (mod d).*

Proof. We may suppose f primitive, at the cost of replacing the congruence condition on $w(S)$ by $w(S) \equiv -1$ (mod dm), for any assigned $m \neq 0$. Let h be prime to dm.

It is clearly possible to find a primitive $\mathbf{t}$ with $f(\mathbf{t}) \neq 0$, prime to dm, and then a primitive $\mathbf{z}$ with $\mathbf{z} \equiv h\mathbf{t}$ (mod dm), $f(\mathbf{t})$ prime to $dmf(\mathbf{z})$, and $f(\mathbf{t})f(\mathbf{z}) < 0$. Now by Theorem 56 the weights of $U(\mathbf{t})$, $U(\mathbf{z})$ are $a^{-1}f(\mathbf{t})$, $a'^{-1}f(\mathbf{z})$ for some a, a' each dividing d, hence each ± 1. This gives $w(U(\mathbf{t})) = |f(\mathbf{t})|$, $w(U(\mathbf{z})) = |f(\mathbf{z})|$. As these integers are coprime by construction, we see by (159) that taking $S = U(\mathbf{t})\, U(\mathbf{z})$ we have, with h such that $hf(t) \equiv 1$,

$$w(S) = |f(\mathbf{t})f(\mathbf{z})| = -f(\mathbf{t})f(\mathbf{z}) \equiv -1 \ (\text{mod } dm).$$

In the next three theorems we study the system of congruences

$$p^{\delta} \| f(\mathbf{t}), \quad p^{\delta} | \mathbf{t}'A; \qquad p^{\eta} \| f(\mathbf{z}), \quad p^{\eta} | \mathbf{z}'A. \tag{192}$$

THEOREM 66. *If p is odd and there exist integral δ, η, $\mathbf{t}$, $\mathbf{z}$ satisfying (192) and*

$$\delta \equiv \eta \ (\text{mod } 2), \quad (p^{-\delta - \eta} f(\mathbf{t}) f(\mathbf{z}) \,|\, p) = -1, \tag{193}$$

then f has a rational automorph S with $w(S)$ prime to d, such that $(w(S) \,|\, p) = -1$ and $w(S)$ is a p'-adic square for every $p' \neq p$ dividing d.

Proof. Here again we may suppose f primitive, at the cost of replacing d by dm ($m \neq 0$) in the conditions on $w(S)$.

The vectors $\mathbf{t}$, $\mathbf{z}$ may be supposed primitive; if $p \,|\, \mathbf{t}$ then $(192)_2$ gives $p^{\delta+1} | \mathbf{t}'A\mathbf{t} = 2f(\mathbf{t})$, contradicting $(192)_1$. We may also suppose them chosen so that $p^{-\delta}f(\mathbf{t})$ and $p^{-\eta}f(\mathbf{z})$ are prime to each other and to dm; and that $f(\mathbf{t})f(\mathbf{z})$ is positive. Further, we may for each $p' \neq p$ dividing dm suppose $f(\mathbf{t})f(\mathbf{z})$, whence also (since $\delta + \eta$ is even) $p^{-\delta-\eta}f(\mathbf{t})f(\mathbf{z})$, to be a p'-adic square.

Now we take $S = U(\mathbf{t})\,U(\mathbf{z})$, and as in Theorem 65 we find $w(S) = p^{-\delta-\eta}f(\mathbf{t})f(\mathbf{z})$, whence the result.

THEOREM 67. (i) *If* (192), *with* $p = 2$, *can be satisfied simultaneously with*

$$\delta \equiv \eta \;(\text{mod } 2), \quad 2^{-\delta-\eta}f(\mathbf{t})f(\mathbf{z}) \equiv -3 \;(\text{mod } 8), \qquad (194)$$

then f *has a rational automorph* S *with* $w(S)$ *prime to* d, $w(S) \equiv -3$ (mod 8), *and* $w(S)$ *a* p-*adic square for every odd* p *dividing* d.

(ii) *If the hypothesis of* (i) *holds good with*

$$\delta \equiv \eta \;(\text{mod } 2), \quad 2^{-\delta-\eta}f(\mathbf{t})f(\mathbf{z}) \equiv -1 \;(\text{mod } 4), \qquad (195)$$

in place of (194), *then the conclusion holds good with* $w(S) \equiv -1$ (mod 4) *in place of* $w(S) \equiv -3$ (mod 8).

Proof. The argument is like that of Theorem 66. There is one slight difference; it may be that 2 divides $\mathbf{t}$ (or $\mathbf{z}$). If so, however, we can put $2^{-1}\mathbf{t}$, $\delta - 2$ for $\mathbf{t}$, δ.

THEOREM 68. *If (for odd* p) *the hypothesis of Theorem 66 fails then* $p^{\frac{1}{2}n(n-1)} \,|\, d$. *If the hypothesis of Theorem 67* (i) *fails then* $8^{\frac{1}{2}n(n-1)} \,|\, d$. *If that of Theorem 67* (ii) *fails then* $4^{\frac{1}{2}n(n-1)} \,|\, d$.

Proof. Suppose first that $p \neq 2$. The simultaneous solubility of (192) and (193) is a property of f which is clearly invariant under p-adic equivalence. Moreover, if f has this property then so too has mf ($m \neq 0$) (with the same $\mathbf{t}$, $\mathbf{z}$ and with δ, η increased by θ, where $p^\theta \,\|\, m$); and also $f+f_1$, for every f_1. (To see this take $t_i = z_i = 0$ for $i > n(f)$.) Next notice that with $\delta = \eta = 0$ (192) and (193) reduce to

$$(f(\mathbf{t})f(\mathbf{z})\,|\,p) = -1, \qquad (196)$$

which is certainly soluble if $n \geqslant 2$ and $p \nmid d$.

Using these results and appealing to Theorem 32 (with a slight change of notation in (105)) we see that if (192) and (193) are not simultaneously soluble we must have

$$f \underset{p}{\sim} a_1 p^{r_1} x_1^2 + \ldots + a_n p^{r_n} x_n^2, \quad p \nmid a_1 \ldots a_n, \qquad (197)$$

with the r_i all unequal, which clearly gives $p^{\frac{1}{2}n(n-1)} \,|\, d$.

Now let $p = 2$; we deal first with (192) and (194). The foregoing remarks leading up to (196) still hold good, except that (196) has to be replaced by $f(\mathbf{t})f(\mathbf{z}) \equiv -3 \pmod 8$. We use also the fact that (192) and (194) are soluble for $f = x_1^2 + 4^r a x_2^2$, a odd, $r = 0, 1$, or 2. (Take $\mathbf{t} = \{1, 0\}$, $\delta = 0$, $\mathbf{z} = \{1, 2\}$, $\{1, 1\}$, or $\{2, 1\}$, $\eta = 0, 0$, or 2.) Combining these results and using Theorem 35, we see that (192) and (194) are soluble unless (197) holds with $p = 2$ and with no two r_i differing by 0, 2 or 4. This gives $r_1, \dots,$ $r_n \geqslant 0, 1, 6, 7, \dots$. Now $2^u \| d$, where, see (10),

$$u = 2[\tfrac{1}{2}n] + r_1 + \dots + r_n = r_1 + (r_2 + 2) + r_3 + \dots,$$

giving $u \geqslant 3n(n-1)/2$.

Now let $p = 2$ and consider (192) and (195). By the arguments for odd p, with (196) replaced by $f(\mathbf{t})f(\mathbf{z}) \equiv -1 \pmod 4$, and using Theorem 35 instead of Theorem 32, we see that (192) and (195) are soluble unless (197) holds (with $p = 2$) and no three of the r_i are equal. We obtain the desired result $2^{n(n-1)} \mid d$ if we can show that (192) and (195) are soluble unless $r_1, \dots, r_n \geqslant 0, 0, 4,$ $4, \dots$. That is, we must solve (192) and (195) if two of the r_i differ by 1 or 3, or if the greatest and least of some three of them differ by 2.

By the remarks at the beginning of this proof, it suffices to consider two cases:

(i) $f = x_1^2 + 2^r a x_2^2$, $r = 1$ or 3, a odd. Here we take $\delta, \eta = 0, r-1$, $\mathbf{t}, \mathbf{z} = \{1, 0\}$, $\{2^{\frac{1}{2}r-\frac{1}{2}}, 1\}$;

(ii) $f = x_1^2 + 4^r a x_1^2 + 4b x_3^2$, ab odd, $r = 0$ or 1. We take $\delta, \eta = 0, 2$, $\mathbf{t} = \{1, 0, 0\}$. One of

$$\mathbf{z} = \{0, 2^{1-r}, 0\}, \quad \{0, 0, 1\}, \quad \{2, 2^{1-r}, 1\}$$

will do what is wanted, since we have $4 \mid \mathbf{z}'A$ in each case and $\tfrac{1}{4}f(\mathbf{t})f(\mathbf{z}) = \tfrac{1}{4}f(\mathbf{z}) = a, b, 1+a+b$.

It would obviously be of interest to consider the consequences of the solubility (or insolubility) of (192) with $\delta \not\equiv \eta \pmod 2$; but the results are more complicated. We therefore state in the form of a lemma just as much, on this point, as will be needed later.

LEMMA. (i) *If* (192) *can be satisfied with* $p = 2$ *and* $\delta + \eta$ *odd, then f has a rational automorph S with $w(S)$ prime to $2d$ and $2w(S)$ a p-adic square for every odd prime p dividing d.*

(ii) *If* (192), *with* $p = 2$, *implies* $\delta \equiv \eta$ (*mod* 2), *then all the r in* (117) *for which the form* $f^{(r)}$ *does not vanish identically must be of the same parity; and every binary form* $\phi^{(r)}$ (*see* (118)) *must be congruent modulo* 4 *to a difference of squares.*

Proof. (i) We argue as in Theorems 66 and 67. Notice that we say nothing about the 2-adic class of $w(S)$; we may of course suppose $w(S) \equiv 1$ (mod 8) if both parts of Theorem 67 apply, which is the only case in which this lemma will be used.

(ii) The assertion regarding the r is obvious. It suffices to prove the second point for the form $x_1^2 + cx_2^2$, $c \equiv 1$ (mod 4). We take $\delta, \eta = 0, 1$ and $\mathbf{t}, \mathbf{z} = \{1, 0\}, \{1, 1\}$.

It will be important later to notice that the automorphs constructed in this section are all products of reflexions each with denominator prime to d.

3. A simpler congruence condition for spinor-relatedness

Denote temporarily by d' the product of the distinct odd primes for which the congruences (192) and (193) cannot be satisfied. Then by Theorem 68 we have $d'^{\frac{1}{2}n(n-1)} \mid d$. By an argument used in the proof of Theorem 32 it can be seen that $p \mid d'$ if and only if (197) holds with unequal r_i and with $(a_i a_j \mid p) = 1$ whenever $r_i \equiv r_j$ (mod 2). Now if (192) and (194) are insoluble, define $d_0 = d_0(f)$ to be $8d'$. If (192) and (194) are soluble but (192) and (195) are not, put $d_0 = 4d'$. Otherwise, $d_0 = d'$. Plainly d_0 is an invariant of the congruence class of f, and we have

$$d_0^{\frac{1}{2}n(n-1)} \mid d. \tag{198}$$

THEOREM 69. *Suppose that* f, f^R *are both integral, with the same discriminant* $d \neq 0$, *prime to* $w(R)$. *Then* f, f^R *are in the same spinor genus if and only if* f *has a rational automorph* S *with* (*for some integer* k)

$$k^2 w(R) w(S) \equiv 1 \; (\text{mod} \, d_0), \quad w(S) \text{ prime to } d. \tag{199}$$

Moreover, the right member of (199)$_1$ *may be replaced by* ± 1 *if* f *is indefinite.*

Proof. By Theorem 64 we need only prove the 'if'. It is simplest to begin by showing that the right member of (190) can be replaced by ± 1 if f is indefinite. We use Theorem 65. If $k^2 w(R) w(S) \equiv -1$ (mod d) and $w(S_1) \equiv -1$ (mod d), then

Theorem 55 gives us that $w(SS_1)\,w(S)\,w(S_1)$ is a square, and so for some k' we have $k'^2w(R)\,w(SS_1) \equiv 1 \pmod{d}$.

Now it is clear from Theorem 64 that a sufficient condition for $f \overset{\cdot}{\sim} f^R$ is

$$k^2 w(R)\,w(S) \equiv 1 \pmod{8p_1p_2\ldots},$$

where p_1, p_2, ... are the distinct odd primes dividing d. An argument like that just used shows that the factor p_i can be left out of the modulus of this congruence if the conclusion of Theorem 66 holds; that is, by the definition of d_0, if $p_i \nmid d_0$. If so, however, we have to retain the condition $p_i \nmid w(S)$, which is implied by (190) but not by (199)$_1$.

We consider similarly whether the 8 in the congruence can be replaced by 4 or omitted, and the result follows. Taking $S = I$ we obtain two corollaries, for the second of which Theorem 63 is needed:

COROLLARY 1. *Two forms f, f' in the same genus, with discriminant d and rank n, are spinor-related if d is not divisible by the $\frac{1}{2}n(n-1)$-th power of any integer greater than 2.*

COROLLARY 2. *Two forms in the same indefinite genus, with discriminant d and rank $n \geqslant 3$, are equivalent if d is not divisible by the $\frac{1}{2}n(n-1)$-th power of any integer greater than 4.*

It is quite easy, for any given form f, to find all the possibilities (prime to d and reduced modulo d) for the weight of a reflexion U of f. Indeed if $U_1 = U(\mathbf{t})$ satisfies g.c.d. $(w(U_1),\,d) = 1$, and $\mathbf{z} \equiv \mathbf{t} \pmod{d^2}$, then it follows from Theorem 56 that $U_2 = U(\mathbf{z})$ has $w(U_2) \equiv \pm w(U_1) \pmod{d}$. Here the ambiguous sign is that of $f(\mathbf{t})f(\mathbf{z})$, and because of the last clause of Theorem 69 it is not important. Hence one can find, for given f, R, whether or not it is possible to satisfy (199) with S a product, say $U_1\ldots U_u$, of reflexions each with weight prime to d. In so doing one may, by Theorem 55, put $w(U_1)\ldots w(U_u)$ for $w(U_1\ldots U_u)$, the product of these numbers being a square.

We shall prove later, with some difficulty, that if (199) cannot be satisfied with such an S then it cannot be satisfied at all.

From Theorem 69 we deduce:

THEOREM 70. *Let f be an integral form and R a rational matrix with $|R| = \pm 1$, $w(R)$ prime to $d(f)$, and f^R integral. Then the* *spinor genus of f^R is uniquely determined by f and $w(R)$. More precisely, if $|R_1|^2 = |R_2|^2 = 1$, and $w(R_1)\,w(R_2)$ is prime to $d(f)$ and congruent to a square modulo $d_0(f)$, then f^{R_1} and f^{R_2} are spinor-related.*

Proof. By Theorem 69, with $f(R_1 \mathbf{x})$, R_1^{-1}, I for f, R, S, it suffices to show that $w(R_1^{-1} R_2)$ is prime to d and congruent to a square modulo d_0. By hypothesis, $w(R_1)\,w(R_2)$, which by Theorem 54 is equal to $w(R_1^{-1})\,w(R_2)$, satisfies these conditions. Since the integral form f^{R_1} is taken into integral forms (f, f^{R_2}) by transformations with matrices R_1^{-1}, $R_1^{-1} R_2$, Theorem 55 tells us that $w(R_1^{-1} R_2)\,w(R_1^{-1})\,w(R_2)$ is a square; the theorem now follows.

Denote by α the number of distinct prime factors of d_0, with the convention that 2 counts twice if $8\,|\,d_0$. Then Theorem 70 shows that the number of spinor genera in the genus of f does not exceed 2^α. The actual number is easily seen to be a power of 2, say 2^β, $\beta \leqslant \alpha$.

4. Decomposition

Given a decomposition

$$f \simeq \phi_1 + \ldots + \phi_\nu \tag{200}$$

of a form f under semi-equivalence (see Theorem 47) we consider whether we can find a decomposition

$$f \mathrel{\dot{\simeq}} f_1 + \ldots + f_\nu, \tag{201}$$

with

$$f_i \simeq \phi_i \quad (i = 1, \ldots, \nu). \tag{202}$$

That is, the new decomposition is to be essentially the same as the old under semi-equivalence, but to be valid also under the narrower relation of spinor-relatedness. We prove:

THEOREM 71. *To every decomposition (200) with at least one summand ϕ_i having rank at least 3 there corresponds a decomposition (201) satisfying (202).*

Remark. It follows at once, by Theorem 47, that every f with $n \geqslant 12$ has a decomposition of the shape (201) with f_i of rank at most 8, 11 for $i <, \leqslant \nu$; and that if $|s(f)| \leqslant n - 4$ then the constants 12, 11 may be replaced by 8, 7.

Proof. Without loss of generality, suppose $n(\phi_1) \geqslant 3$. Write (200), using Theorem 50, as

$$f^R = \phi_1 + \ldots + \phi_\nu, \quad w(R) \text{ prime to } d(f).$$

Choose a prime p (using Dirichlet's theorem on primes in a progression) with $p \nmid w(R)$, $pw(R) \equiv 1 \pmod{d}$, $d = d(f)$. Clearly $p \nmid d(\phi_1)$ since $d(\phi_1) \mid d$. Hence ϕ_1 is a p-adic zero form (by (95) and Theorem 24). So we may suppose that p^2 divides the leading coefficient of ϕ_1, and that ϕ_1 has no terms in $x_1 x_j$ $(2 < j \leqslant n(\phi_1))$ (see Theorem 12). This means that

$$\phi_1(p^{-1}x_1, px_2, x_3, \ldots)$$

is an integral form; call it ϕ_1'.

Now we have, with $P = [p^{-1}, p, 1, \ldots, 1]$,

$$f^{RP} = \phi_1' + \phi_2 + \ldots + \phi_\nu,$$

with $\phi_1' \simeq \phi_1$ by Theorem 50 and $p \nmid d(\phi_1)$. If we can prove $f^{RP} \stackrel{.}{\sim} f$ this gives what is required (with equality in all but one of the relations (202)). Our construction, however, makes it trivial that $w(RP) = pw(R) \equiv 1 \pmod{d}$; hence (190) can be satisfied with $S = I$ and $f^{RP} \stackrel{.}{\sim} f$ follows; and this completes the proof, by Theorem 64.

By Theorem 63, Theorem 71 holds for indefinite f with equivalence for spinor-relatedness in (201). This result has, however, no analogue for definite forms. To see this, let f be positive-definite and perfect (Chapter 2, §11). Then f is not decomposable under equivalence; for it follows at once from the definition that (i) a perfect form is not disjoint and (ii) perfectness is invariant under equivalence.

5. Representation of integers by a spinor genus

It is convenient to say that the spinor genus of f represents the integer a (properly) if some form in the spinor genus represents a (properly); similarly with *genus* or *class* for *spinor genus*. Representation by the class of f is of course the same as representation by f itself. We shall generally consider only proper representation; corresponding results when improper representation is admitted follow at once.

8

We shall first show that *if $n(f) \geqslant 4$ then the spinor genus of f represents a properly if and only if the genus of f does so.* The 'only if' is trivial, and we have already (Theorem 53) proved the 'if' for indefinite forms. It suffices, therefore, to prove the italicized result for positive forms (though in the proof below we use the assumption that f is definite only to exclude the (trivial) case $a = 0$. And by Theorem 51 we may restate what needs proof as

THEOREM 72. *If f is a positive form of rank $n \geqslant 4$, with discriminant d, and a is a positive integer such that the congruence $f(\mathbf{x}) \equiv a \pmod{d}$ has a solution with $\mathbf{x}$ integral and primitive, then some form f' in the spinor genus of f represents a properly.*

Proof. By Theorem 51, there is a rational R with $w(R)$ prime to d and $|R| = \pm 1$, such that f^R has leading coefficient a. We choose a prime p with $p \nmid 2adw(R)$, $pw(R) \equiv 1 \pmod{d}$. By a suitable parallel transformation (with integral coefficients) we may suppose that the coefficients of $x_1 x_2, ..., x_1 x_n$ in f^R are divisible by p; that is

$$f^R = ax_1^2 + ... \equiv ax_1^2 + g(x_2, ..., x_n) \pmod{p},$$

for some g which clearly has $p \nmid d(g)$. The argument of Theorem 71 (using $n(g) = n - 1 \geqslant 3$) now shows that g^P is integral for a P with weight p. Clearly therefore $f^{R[1, P]}$ is integral, and has leading coefficient a.

It remains only to prove that $f^{R[1, P]} \dotdiv f$. As in the proof of Theorem 71, we have $w(R[1, P]) = pw(R) \equiv 1 \pmod{d}$.

Clearly f itself must represent a properly if its spinor genus contains only one class. This is of course trivial if the genus also contains only one class (that is, if the genus contains only one spinor genus); for then the weaker Theorem 51 will also tell us that f represents a properly under the hypotheses of Theorem 72. The simplest instance of a form whose spinor genus contains only one class though its genus contains more than one spinor genus is

$$x_1^2 + x_1 x_2 + 7x_2^2 + 3x_3^2 + 3x_3 x_4 + 3x_4^2,$$

with $n = 4$, $d = 3^6$. This form represents properly every a with $a > 0$, $81 \nmid a$, $a \not\equiv -1 \pmod 3$, $a \not\equiv -3 \pmod 9$. We cannot yet prove the assertion that the spinor genus of this form contains one class only.

It is clear from (198) and Theorem 69 that the foregoing example is the simplest of its kind. It would be more interesting to have similar results for $n = 3$; these would be new in the indefinite case too. Unfortunately, when $n = 3$ the spinor genus may not represent properly all the integers so represented by the genus. As an example, take

$$f = x_1^2 + x_1 x_2 + x_2^2 + 9x_3^2, \quad n = 3, \, d = -27.$$

We shall show that f does not represent $4h^2$ properly if h is positive and congruent to 1 (mod 3).

To prove this, write $f = 4h^2$ as

$$\psi(x_1, x_2) = (2h - 3x_3)(2h + 3x_3), \tag{203}$$

where $\psi = x_1^2 + x_1 x_2 + x_2^2$. Both factors on the right of (203) are positive, and congruent to -1 modulo 3. Hence there is a prime $p \equiv -1$ (mod 3) such that, for some odd integer $2u - 1$, we have $p^{2u-1} \| 2h - 3x_3$. Now either $p \mid 2h + 3x_3$, giving $p \mid 2h$, $p \mid x_3$, or $p^{2u-1} \| \psi$. Assuming the solution of (203) to be proper, this gives

$$p \mid \psi, \quad p \nmid (x_1, x_2); \quad \text{or} \quad p^{2u-1} \| \psi.$$

These are both impossible. For $d(\psi) = -3$ is not a p-adic square and so $p \mid \psi$ implies $x_1 \equiv x_2 \equiv 0$ (mod p), $p^2 \mid \psi$; repeating the argument, $p^{2u-1} \mid \psi$ implies $p^{2u} \mid \psi$.

It is easily shown by the methods of Chapter 2 that the only other class in the genus of f is represented by

$$f' = x_1^2 + 3(x_2^2 + x_2 x_3 + x_3^2).$$

A similar argument shows that f' will not represent $4h^2$ properly if $h > 0, h \equiv -1$ (mod 3). (One of the positive integers $2h \pm x_1$ has to be congruent to 0, and the other therefore to $4h \equiv -1$, modulo 3.)

To see that this behaviour is anomalous from the present point of view we need to show that f, f' are in different spinor genera; or, equivalently, that the genus of f contains two spinor genera. Looking at (190), where clearly $w(R) = 2$ is possible, we see that we must show that $w(S)$ prime to 3, $f^S = S$, implies $w(S) \equiv 1$

(mod 3). This is easily seen to be true for reflexions (the integer a of Theorem 56 must be one of 1, 3, 9, 27; but $27 \mid \mathbf{t}'A$ implies $81 \mid f(\mathbf{t})$, while with $a = 1, 3$, or 9 we find always that $a \mid \mathbf{t}'A$ implies $a^{-1}f(\mathbf{t}) \equiv 1 \pmod{3}$). The desired result follows generally when we show later that S in (190) may be taken to be a product of reflexions each with denominator prime to d.

The two indefinite forms (with $d = 125$)

$$\phi + 25x_3^2, \quad x_1^2 + 5\phi; \qquad \phi(x_1, x_2) = x_1^2 + x_1 x_2 - x_2^2,$$

show similar anomalous behaviour. If h is prime to 5, the first of these forms represents $4h^2$ properly only if $h \equiv \pm 2 \pmod 5$, while the second does so only if $h \equiv \pm 1 \pmod 5$. The argument is similar but a little simpler, since by Theorem 63 the fact that the two forms are in different spinor genera follows from the obvious fact that they are inequivalent.

6. The exceptional integers of a ternary genus

We shall say that the integer a is EXCEPTIONAL, in relation to the genus of a ternary form f, if some form in the genus represents a properly but some spinor genus in the genus fails to do so. We shall carry the investigation of the exceptional integers far enough to show that they are appropriately so called, and can all be found in any numerical example. The problem of representation of integers by indefinite ternary forms will then be essentially solved. That of representation by definite ternaries will have been carried a stage further. For example, we shall see that all the exceptional integers of the genus with $n = s = 3$, $d = -27$ discussed in the last section are squares. These are easily dealt with separately; and each of the two forms (being in the only class in its spinor genus) represents properly every non-square integer satisfying the generic conditions.

In choosing a rational matrix P, with weight $p \nmid d = d(f)$, such that, for given f, f^P will be in the spinor genus we wish, it is convenient to assume

$$f \equiv x_1 \dot{x}_2 + dx_3^2 \pmod{p^2}. \tag{204}$$

This is permissible, by an equivalence transformation. For Theorem 29, with $p \nmid d$, shows that $f^M \equiv x_1 x_2 + dx_3^2 \pmod{p^2}$ for some integral M, clearly with $|M|^2 \equiv 1 \pmod{p^2}$, and the argument of Theorem 41 shows that we can choose M with $|M| = \pm 1$. It is also convenient to take P to be a reflexion of the form on the right of (204):

$$P = I - p^{-1}\mathbf{t}\mathbf{t}'\begin{pmatrix} . & 1 & . \\ 1 & . & . \\ . & . & 2d \end{pmatrix}, \quad \text{where} \quad t_1 t_2 + dt_3^2 = p, \quad (205)$$

and $\mathbf{t} = \{t_1, t_2, t_3\}$. Thus $P^2 = I$, $(x_1 x_2 + dx_3^2)^P = x_1 x_2 + dx_3^2$, and (204) shows that f^P is integral.

The spinor genus of f^P, constructed in this way, is determined by f and the residue of p modulo $d_0 = d_0(f)$, defined in §3. More precisely, if two forms f, f_1 in the same genus are transformed in this way (with the same prime p) into $f^P, f_1^{P_1}$, then $f^P, f_1^{P_1}$ are in the same spinor genus if and only if f, f_1 are so. For, taking f_1, by Theorem 50, to be f^R, $w(R)$ prime to pd, f^P and $f_1^{P_1}$ are related by $R_1 = P^{-1}RP_1$, with $w(R_1)$ either $w(R)$ or $p^2w(R)$ by Theorem 55. Now we appeal to Theorem 64 and note that any automorph of f^P, with weight prime to d, can be expressed as $P^{-1}SP$, $w(S)$ prime to d, $w(P^{-1}SP) = w(S)$ or $p^2w(S)$.

THEOREM 73. *If a is exceptional in relation to the genus of f, then $a \neq 0$; and if we write*

$$ad = a_0 a_1^2, \quad a_0 \text{ square-free, } a_1 \text{ integral}, \quad (206)$$

where $d = d(f)$, then $a_0 \neq 1$ and we have

$$\prod_{p|2a_0} (a_0, w(S))_p = 1 \quad (207)$$

for every S with $f^S = f$ and

$$\text{g.c.d. } (2da_0, w(S)) = 1. \quad (208)$$

Proof. We may suppose that f itself represents a properly, say $f(\mathbf{y}) = a$, where $\mathbf{y}$ is integral and primitive. We may also suppose $a \neq 0$; for we already know that representation of 0 is a property of the genus. Now suppose (Theorem 50) that the spinor genus of f^R, with $w(R)$ prime to $2ad$, fails to represent a properly. Choose a prime p' with $p' \equiv w(R)$ (mod $8ad$).

We know that $f^P \stackrel{\cdot}{\sim} f^R$ if f^P is integral and $w(P) = p'$ (for $w(P^{-1}R) = p'w(R)$ or $p'^{-1}w(R)$, by Theorem 55, is congruent to a square modulo d and we appeal to Theorem 64 with

f^P, I for f, S). Now choose P so that, further, $P^2 = 1$. Then if $P\mathbf{y}$ is integral we have $f^P(P\mathbf{y}) = f(\mathbf{y}) = a$, and $P\mathbf{y}$ is primitive since $p' \nmid a$. Hence by hypothesis $P\mathbf{y}$ cannot be integral.

This means, referring to (204) and (205) with p' for p, that there are no integers t_1, t_2, t_3 with

$$t_1 t_2 + dt_3^2 = p', \quad p' \, | \, \mathbf{t}' \begin{pmatrix} \cdot & 1 & \cdot \\ 1 & \cdot & \cdot \\ \cdot & \cdot & 2d \end{pmatrix} \mathbf{y} = t_2 y_1 + t_1 y_2 + 2dt_3 y_3.$$

Putting $t_1 = 1$, $t_2 = p' - dt_3^2$, this means that the congruence

$$-dt_3^2 y_1 + 2dt_3 y_3 + y_2 \equiv 0 \pmod{p'}$$

does not hold for any integer t_3. Hence either $p' \,|\, y_1, y_3$, which makes $f(\mathbf{y}) = a \equiv 0 \pmod{p'}$ contrary to our choice of p', or $4d^2 y_3^2 + 4dy_1 y_2 \equiv 4ad \pmod{p'}$ is a quadratic non-residue modulo p'. That is, $(a_0 | p') = -1$.

We write this, using (83), as $\Pi(a_0, p')_p = -1$, the product being over $p \,|\, 2a_0$. By our choice of p' it follows at once that

$$\prod_{p | 2a_0} (a_0, w(R))_p = -1, \quad \text{if} \quad w(R) \text{ is prime to } 2ad.$$

Plainly this formula cannot hold if $a_0 = 1$; and it must still hold if we put $w(SR)$ for $w(R)$, S any automorph of f with $w(S)$ prime to $2ad$. Further, using Theorem 55 and (77), we may put $w(S)\,w(R)$ for $w(R)$. But now using the multiplicative property (81) of the Hilbert symbol, we see that (207) must be implied by g.c.d. $(w(S), 2ad) = 1$.

It remains only to show that the weaker (208) implies (207). By Theorem 60, we can vary our choice of S so as to leave the residue of norm S unaltered modulo $8a_0 d$, thus not affecting (207), and at the same time get rid of any factors dividing $2ad$ but not $2a_0 d$.

THEOREM 74. *With the hypothesis of Theorem 73 we have*

$$a_0 \,|\, d_0, \quad a_0 \equiv 1 \pmod 4 \quad \text{if} \quad 2 \nmid d_0, \tag{209}$$

for $d_0 = d_0(f)$ defined in §3, also $a_0 \neq 1$, always, and $a_0 > 1$ if f is indefinite.

Proof. Suppose first that an odd prime p_1 divides a_0 but not d_0. Choose S, with $w(S)$ prime to $2a_0d$, so that $w(S)$ is a p-adic square for every p dividing $2a_0d$, except p_1, while $(w(S)|p_1) = -1$. Theorem 66 shows that this is possible. Then every factor on the left of (207) is 1, except that with $p = p_1$, and since a_0 is square-free by definition this factor, with $p_1 \| a_0$, is $(w(S)|p_1)$ by (78). Thus (207) fails while (208) holds. This contradiction gives us $a_0 | 2d_0$.

Now suppose $2 \nmid d_0$. We choose S with $w(S)$ prime to $2a_0d$ so that $w(S)$ is a p-adic square for every odd p dividing $2a_0d$. At the same time we can have $w(S) \equiv -3 \pmod 8$ or $-1 \pmod 4$, as we choose, by Theorem 67. Either choice reduces (207) to

$$(a_0, w(S))_2 = 1,$$

which is false by the first choice if $2 \| a_0$, and by the second if $a_0 \equiv -1 \pmod 4$.

As we have already noted $a_0 \neq 1$, it remains only to prove $a_0 > 0$ if f is indefinite. We use Theorem 65 to find an S satisfying (208) and with $w(S) \equiv -1 \pmod d$. Suppose for the moment that $w(S) \equiv -1 \pmod 8$. Then (207) gives (using (209), which gives $a_0 | d$) $\Pi(a_0, -1)_p = 1$. The product, which is over $p | 2a_0$, is $(a_0, -1)_\infty = \operatorname{sgn} a_0$ by (83).

Now if our choice of S with $w(S) \equiv -1 \pmod d$ does not give $w(S) \equiv -1 \pmod 8$ then we have $8 \nmid d$, giving $8 \nmid d_0$ by (198). If $4 | d$ we have $w(S) \equiv -1 \pmod 4$ and may suppose $w(S) \equiv -1 \pmod 8$ by using Theorem 67. If $4 \nmid d$ then d_0 is odd and again Theorem 67 permits us to suppose $w(S) \equiv -1 \pmod 8$. These remarks complete the proof.

THEOREM 75. *If a is exceptional in relation to the genus of f, and $p \nmid d(f)$, then ap^2 is also exceptional; and the converse is true unless $p = 2$ and a is odd.*

Proof. Suppose that f represents ap^2 properly, say $f(\mathbf{y}) = ap^2$, $\mathbf{y}$ primitive. Assume (204) and consider f^P, with a suitably chosen P (see (205)). We choose P so that $P\mathbf{y} \equiv 0 \pmod p$. If so, $p^{-1}P\mathbf{y}$ is clearly primitive, and so f^P represents a properly. If this construction can be carried out for some f in each spinor genus it gives an f^P in each spinor genus, and so proves that if ap^2 is not exceptional then a cannot be.

To see that the construction is possible, note that by the argument of Theorem 73 it suffices to solve

$$-dy_1t_3^2 + 2dy_3t_3 + y_2 \equiv 0 \pmod{p}.$$

With $\qquad p \,|\, f(\mathbf{y}) \equiv y_1y_2 + dy_3^2 \pmod{p}$

this congruence is trivially soluble, reducing to $(y_1t_3 - y_3)^2 \equiv 0$ $\pmod{p}$, unless $p\,|\,y_1$, $p\,|\,y_3$, $p\nmid y_2$. This case is easily dealt with by putting $\mathbf{t} = \{p, 1, 0\}$.

To prove the converse, suppose $f(\mathbf{y}) = a$, $\mathbf{y}$ primitive. It suffices to choose P so that $P\mathbf{y}$ is not integral. If so, then $pP\mathbf{y}$ is clearly integral and primitive, and so f^P represents ap^2 properly. If this construction can be carried out for one f in each spinor genus, then in each spinor genus there is an f^P representing ap^2 properly, and ap^2 is not exceptional.

This second construction reduces to choosing $\mathbf{t}$ with

$$t_1t_2 + dt_3^2 = p$$

so that $\qquad p \nmid t_2y_1 + t_1y_2 + 2dt_3y_3.$

If it cannot be done by taking $\mathbf{t} = \{1, p, 0\}$ or $\{p, 1, 0\}$ then we have $p\,|\,y_1, y_2$. With a even and $p = 2$, this makes $2\,|\,\mathbf{y}$, so we may suppose $p \neq 2$. We put $t_1 = t_3 = 1$ and $t_2 = p - d$.

These three theorems show that there are infinitely many exceptional integers if there are any at all, but that it suffices to consider those which have no prime power factor prime to d except possibly 4. It is not too difficult to find whether any given a is exceptional; we construct all possible forms f given by

$$4af = (2ax_1 + \ldots)^2 + g(x_2, x_3), \quad g \text{ reduced,}$$

and see if there is a form in the genus not spinor-related to any of them.

The problem of finding all the exceptional integers would thus be reduced to a finite one if we could deal with the possibility $p^u\,|\,a$, u large, $p\,|\,d$. This arises only if f is a p-adic zero form. If so, then Theorem 39 gives $f^M = p^rf_0$, for some integral M with $|M|$ a power of p, and some r, and an f_0 with $p\nmid d(f_0)$, $d(f)/d(f_0)$ a power of p. Thus it suffices to solve $f(M\mathbf{y}) = ap^{-r}$, with $\mathbf{y}$ primitive and $p\nmid M\mathbf{y}$. This supplementary condition does not give much trouble in any numerical example, and drops out if improper representations are allowed. We shall therefore not go into the problem in detail, but merely remark that it reduces to the study of a simpler form.

7. Representation of integers by positive forms

There are various inductive methods of tackling this problem for forms of rank not less than 4. These could be used to prove:

*THEOREM 76. *Let f be a positive form of rank at least* 4, *and a an integer represented properly by the genus of f. Then f represents a properly if a is sufficiently large.*

This theorem can be deduced from results in the literature, obtained by analytical methods. By 'sufficiently large' is meant 'exceeding a bound depending only on f'. The example

$$a = 2b - 1, \quad f = 2(x_1^2 + \ldots + x_{n-1}^2) + (2b+1)\, x_n^2,$$

b a large positive integer, shows that this bound may be large with $|d|$.

The word 'properly' may be omitted each time if $n \geqslant 5$, but not if $n = 4$. To see the latter point, take

$$f = x_1^2 + x_2^2 + 7x_3^2 + 7x_4^2, \quad a = 3 \cdot 7^{2u}, \quad u \text{ large}.$$

It is easily seen that a solution of $f = a$ would have $7^u \,|\, \mathbf{x}$, and on cancelling 7^u we should deduce a solution of $f = 3$. This is impossible; but $f \equiv 3 \pmod{d}$ $(d = 2^4 \cdot 7^2)$ is soluble, so the genus of f represents 3 and therefore also $3 \cdot 7^{2u}$.

The problem of representation by positive ternary forms cannot be tackled by induction on n, because binary forms lack the properties (in particular, that of representing arithmetic progressions) which would be needed. There is, however, sometimes a possibility of using induction on $|d|$. Consider, for example, the two forms

$$f = 5x_1^2 + 10x_2^2 + 5x_2x_3 + x_3^2 = 5^{-1}f'(5x_1, 5x_2, x_3),$$
$$f' = x_1^2 + 2x_2^2 + 5x_2x_3 + 5x_3^2 = 5^{-1}f(x_1, x_2, 5x_3),$$

with discriminants -75, -15. Let a be an integer prime to 5. If f' represents $5a$, necessarily with $5 \,|\, x_1$, $x_2 = 5x_1'$, $5x_2'$, say, then f represents a, for $f(x_1', x_2', x_3) = a$. Similarly, f represents $5a$, necessarily with $5 \,|\, x_3, = 5x_3'$, if and only if $f' = a$ is soluble. Thus the problem of representation by f is reduced to that of representation by f' (except for multiples of 25, which cannot be represented properly by either form).

Sometimes one can show by elementary arguments that a form f represents all the integers represented by some other form f' in its spinor genus. If there are just two classes in the spinor genus it follows that f represents all the integers represented by the spinor genus. For an example, take

$$f = x_1^2 + 2x_2^2 + x_2 x_3 + 2x_3^2, \quad f' = x_1^2 + x_1 x_2 + x_2^2 + 5x_3^2;$$

this f is equivalent to the f' of the preceding example. A simple rational relation between these forms is given by

$$f' = f(x_1 + \tfrac{1}{2}x_2, \; x_3 - \tfrac{1}{2}x_2, \; x_3 + \tfrac{1}{2}x_2),$$

which shows that from a solution of $f'(\mathbf{x}) = a$ with x_2 even (or with x_1 even) one can deduce a solution of $f = a$. Since

$$f'(\mathbf{x}) = f'(-x_1 - x_2, \; x_2, \; x_3),$$

a similar use may be made of a solution of $f'(\mathbf{x}) = a$ with x_1, x_2 both odd, and so $-x_1 - x_2$ even. The argument depends on the integral automorphs of f'; it cannot be reversed because f has no useful integral automorph. Indeed, f' fails to represent 2, although $f = 2$ is soluble.

We shall take this difficult problem no further.

CHAPTER 8

THE GENERAL RATIONAL AUTOMORPH

1. Factorization of automorphs into reflexions

The general rational automorph S of a form f is not easy, at least for $n \geqslant 4$, to handle directly. It is best dealt with by factorizing it into reflexions. We begin by proving a theorem which, apart from some complications involving the prime number 2, will fill the gap, mentioned in the remark preceding Theorem 70, in the theory of the last chapter.

THEOREM 77. *Let m be any odd integer, and S a rational automorph of a form f with $w(S)$ prime to m. Then S is expressible as a product of rational reflexions of f, each with denominator prime to m.*

Proof. Plainly we may suppose f primitive. We may therefore suppose a_{11} prime to m, as we may make an equivalence transformation without affecting the hypothesis or conclusion of the theorem. More generally, we may replace f by f^R if f^R is integral, $|R| = \pm 1$, and $w(R)$ prime to m. If we do this we must of course replace S by $R^{-1}SR$, and each of the reflexions U by $R^{-1}UR$ (see (172)).

Now, completing the square, write

$$4a_{11}f = (2a_{11}x_1 + a_{12}x_2 + \dots)^2 + g(x_2, \dots, x_n).$$

The theorem is true for f if it is true for $4a_{11}f$. Transforming $4a_{11}f$ in the obvious way, we get rid of the terms in $x_1x_2, \dots, x_1x_n$; the denominator of the matrix R used is a divisor of $2a_{11}$, and prime to m. Changing the notation for simplicity, this means that we may suppose

$$f = a_{11}x_1^2 + g(x_2, \dots, x_n), \quad a_{11} \text{ prime to } m. \tag{210}$$

Write for brevity

$$\mathbf{y} = \{1, \mathbf{0}\}, \mathbf{z} = \{0, 1, \mathbf{0}\}, \quad \text{whence} \quad \mathbf{y}'A\mathbf{z} = 0, \tag{211}$$

by (210). We may suppose, by a suitable equivalence transformation of g, that

$$U(\mathbf{z}, f) = [1, U(\{1, \mathbf{0}\}, g)] \tag{212}$$

(this equality follows easily from (25) and (210)) has denominator prime to m. And we may choose an r such that

$$p^r \nmid f(\mathbf{z}) = g(1, 0, ..., 0), \quad \text{for} \quad p \mid m. \tag{213}$$

We now introduce induction from $n - 1$ to n. The theorem is trivial for $n = 1$ ($S = \pm I$, and $-I$ is the only reflexion). The inductive hypothesis tells us that the theorem is true for S of the shape $[1, S_1]$, where S_1 is necessarily an automorph of g. (If U_1 is a reflexion of g then $[1, U_1]$ is a reflexion of f, cf. (212).) In the notation of (211), this means that the theorem is true for S with $S\mathbf{y} = \mathbf{y}$, $\mathbf{y}'S = \mathbf{y}'$. If, however, the first of these conditions holds, that is, if S has $\mathbf{y}$ as its first column, then (210) shows that $f^S = f$ is impossible unless also $\mathbf{y}'S = \mathbf{y}'$, i.e. $\mathbf{y}'$ is the first row of $\mathbf{S}$. So the theorem is true if $S\mathbf{y} = \mathbf{y}$.

Next, we note that the theorem is true for $S = U_1(U_1 S U)U$ if it is true for $U_1 S U$, U, U_1 being any reflexions of f with denominators prime to m. It follows therefore that the theorem is true if such reflexions U, U_1 can be chosen in such a way that $U_1 S U \mathbf{y} = \mathbf{y}$.

The argument of Theorem 5 shows that this last condition holds if $U_1 = U(\mathbf{y} - SU\mathbf{y})$. In order that U_1 may be defined, and have denominator prime to m, it is sufficient that

$$p \nmid f(\mathbf{y} - SU\mathbf{y}) = 2f(\mathbf{y}) - \mathbf{y}'ASU\mathbf{y} \quad \text{for} \quad p \mid m. \tag{214}$$

We choose $U = U(\mathbf{t})$, with a $\mathbf{t}$ such that

$$\mathbf{t} \equiv \mathbf{y} \pmod{p} \quad \text{or} \quad \mathbf{t} \equiv \mathbf{z} \pmod{p^r} \quad \text{for} \quad p \mid m. \tag{215}$$

Here the choice between the two alternatives is to be made independently for each p dividing m. It is clear, using (213), that however the choice is made $f(\mathbf{t}) \neq 0$, that is $U(\mathbf{t})$ exists, and moreover $U(\mathbf{t}) \equiv U(\mathbf{y})$ or $U(\mathbf{z})$ (mod p). Now we have

$$U(\mathbf{y})\mathbf{y} = -\mathbf{y} \quad \text{and} \quad U(\mathbf{z})\mathbf{y} = \mathbf{y}$$

by $(211)_3$ and (173). So

$$U\mathbf{y} = U(\mathbf{t})\mathbf{y} \equiv \mp \mathbf{y} \pmod{p} \quad \text{for} \quad p \mid m, \tag{216}$$

according as the first or second choice is made in (215).

Now to complete the proof we have only to notice that (214) reduces, by (216), to $p \nmid 2f(\mathbf{y}) \pm \mathbf{y}'AS\mathbf{y}$, for either choice of the sign. If neither choice gave what is wanted we should have $p \mid 4f(\mathbf{y}) = 4a_{11}$, contradicting $(210)_2$.

2. Factorization into reflexions with odd denominators

We prove

THEOREM 78. *The conclusion of Theorem 77 holds for even* $m \neq 0$, *if* f *is 2-adically equivalent to a diagonal form in which no three of the diagonal coefficients are divisible exactly by the same power of* 2.

Proof. The hypothesis implies, by the argument of Theorem 41, that f is equivalent to a form which is congruent, modulo any prescribed power of 2, to a diagonal form with the stated property. We may therefore suppose that f itself is congruent modulo a suitable power of 2 to such a diagonal form. This means that (if, as we may suppose, f is primitive) we have

$$A(f) \equiv [2, 2a_{22}, 0, ..., 0] \pmod 4, \tag{217}$$

and also that g, constructed as in the proof of Theorem 77, will have this property after removal of its divisor if any.

Now we examine the argument of Theorem 77 and show that with the present additional hypothesis it works for even m. First, the terms in $x_1 x_2, ..., x_1 x_n$ can by (217) be got rid of without using a transformation with even denominator; so the assumption (210) remains permissible. Next, the use of the inductive hypothesis still holds good by the remark following (217). It is also necessary to work with a higher power of 2 than of the odd prime p in some congruences; but this is trivial.

It remains therefore only to show that with the notation used in the proof of Theorem 77 one of the reflexions $U(\mathbf{y} \pm S\mathbf{y})$ exists and has an odd denominator. That is, we replace (214), for $p = 2$, by

$$2^r \| f(\mathbf{y} - SU\mathbf{y}), \quad 2^r \,|\, (\mathbf{y} - SU\mathbf{y})\, A,$$

for suitable r. This clearly makes the denominator of $U(\mathbf{y} - SU\mathbf{y})$ a divisor of the odd integer $2^{-r} f(\mathbf{y} - SU\mathbf{y})$, and it is equivalent to

$$2^r \| 2f(\mathbf{y}) \pm \mathbf{y}'AS\mathbf{y}, \quad 2^r \,|\, (\mathbf{y} \pm S\mathbf{y})'A. \tag{218}$$

Since $2 \,|\, A$, $(218)_1$ is impossible with $r = 0$; but it holds with $r \leqslant 2$ by suitable choice of the sign since $4f(\mathbf{y}) = 4a_{11} \not\equiv 0 \pmod 8$. Choosing the sign accordingly, we note that $2 \,|\, A$ makes $(218)_2$

vacuous for $r = 1$, and independent of the choice of sign if $r = 2$. So we have to show that either

$$4 \,|\, \mathbf{y}'AS\mathbf{y} \quad \text{or} \quad 4 \,|\, (\mathbf{y}+S\mathbf{y})'A. \tag{219}$$

(In case $(219)_1$, $(218)_1$ holds for $r = 1$ with either sign.)

Using (217) and writing $S\mathbf{y} = \{\eta_1, ..., \eta_n\}$, failure of $(219)_1$ makes η_1 odd. Then if $(219)_2$ also fails $a_{22}\eta_2$ is also odd. Now using (217) again we have the contradiction

$$1 \equiv f(\mathbf{y}) = f(S\mathbf{y}) \equiv \eta_1 + a_{22}\eta_2 \equiv 0 \pmod{2}.$$

This completes the proof.

It is worth while to show by an example that Theorem 77 does not hold for even m (in particular, for $m = d$ when d is even) for all f. For if it did, we could find the possibilities for norm S, when $w(S)$ is prime to d, very easily, whereas as it is we have to go a long way round to find them. The example is

$$f = x_1^2 + x_1 x_2 + x_2^2 + x_3^2 + x_3 x_4 + x_4^2 + 2x_5^2, \quad d = 18.$$

For this f, Theorem 56 shows that with primitive $\mathbf{t}$ the denominator of $U(\mathbf{t}, f)$ cannot be odd unless $f(\mathbf{t})$ is odd or congruent to 2 modulo 4. In the first case both of t_1, t_2, but not both of t_3, t_4, are even, or vice versa. In the second, to make the denominator of $U(\mathbf{t})$ odd we have to have $2 \,|\, \mathbf{t}'A$, which makes $2 \,|\, t_1, ..., t_4$. A simple calculation now shows that the (i, j) element of $U(\mathbf{t})$ is even (that is, is a rational fraction with even numerator and odd denominator) for $i \leqslant 4, j = 5$, for $i \leqslant 2 < j$, and also for $j = 1$ or 2 and $i = 3$ or 4. Now this property is multiplicative; and there is an S, defined by $S\mathbf{x} = \{x_3, x_4, x_1, x_2, x_5\}$, which does not possess it, although its denominator ($= 1$) is prime to d.

If we are satisfied to prove the falsity of Theorem 77 for even $m \neq 0$, not necessarily $= d$, then an example with $n = 4$ may be found by putting $x_5 = 0$ in the example just dealt with; the proof is similar.

3. Factorization of automorphs; further properties

The factorization obtained in Theorem 77 is not unique. The parity of the number of factors is, however, determined by the sign of $|S|$. Another invariant property of the factorization is given by

THEOREM 79. *Suppose that the rational automorph S of f has the factorization*
$$S = U(\mathbf{t}_1, f) \ldots U(\mathbf{t}_u, f), \tag{220}$$

and let q be the square-free integer such that $q|f(\mathbf{t}_1)\ldots f(\mathbf{t}_u)|$ is a square. Then q depends only on f and S.

Proof. The hypothesis and conclusion are both unaffected if we transform f rationally into f^R, replacing S by $R^{-1}SR$ and each U by $R^{-1}UR$ (see (172)); also if we replace any $\mathbf{t}_\nu$ by $\theta\mathbf{t}_\nu$, θ rational, $\neq 0$. We may therefore suppose the $\mathbf{t}_\nu$ integral and primitive.

If the theorem is false for f it is false also for the $2n$-ary form $f-f$, and for an automorph of this form of the special shape $[S_n, I_n]$, S_n an automorph of f. To see this, note that to a reflexion $U(\mathbf{t}, f)$ of f there corresponds a reflexion $U(\mathbf{z}, f-f)$, $\mathbf{z} = \{\mathbf{t}, \mathbf{0}\}$, of $f-f$ which reduces, using (25), to $[U(\mathbf{t}, f), I_n]$.

Now the form $f-f$ is rationally related to a form (see Theorem 3) of the shape $a_{11}(x_1^2 - x_{n+1}^2) + \ldots + a_{nn}(x_n^2 - x_{2n}^2)$, hence clearly to $x_1 x_2 + \ldots + x_{2n-1} x_{2n}$, which has discriminant 1. It follows therefore that it suffices to prove the theorem for the case $d(f) = 1$, $\mathbf{t}_1, \ldots, \mathbf{t}_u$ integral and primitive.

With this, the integer a of Theorem 56 is necessarily ± 1, and $w(U(\mathbf{t}_\nu, f)) = |f(\mathbf{t}_\nu)|$. Now Theorem 55 shows that $qw(S)$ is a square, which proves the theorem.

Some information about the q of Theorem 79 is given by

THEOREM 80. *Suppose $p \nmid w(S)$, (220) holds, and that*
$$p^\delta \| f(\mathbf{t}), \quad p^\delta | \mathbf{t}'A \text{ together imply } 2 | \delta. \tag{221}$$

Then $f(\mathbf{t}_1)\ldots f(\mathbf{t}_u)$ is divisible exactly by an even power of p.

Proof. We shall prove, and use later, only the case $p = 2$ of this theorem. The proof for $p \neq 2$ is similar to that for $p = 2$, but simpler. It may be remarked that if we replace $2 | \delta$ in (221) by $2 \nmid \delta$, in which case $\delta \neq 0$ makes f imprimitive, then the conclusion still holds if u is even, that is, if $|S| = 1$, but there is a complication if $|S| = -1$.

The theorem (with $p = 2$) is true by the corollary to Theorem 55 if $d = d(f)$ is odd. We shall prove it in general by using rational transformations which take f into a multiple of a form with odd discriminant, without disturbing the hypothesis that $w(S)$ is odd.

Such transformations cannot be found in the general case. We therefore begin by observing that the conclusion of the theorem holds for f if there is any f_1 such that it holds for $f+f_1$. We also note that (221) implies the hypothesis of part (ii) of the lemma proved in Chapter 7, §2. Using the conclusion of that lemma (part (ii)), and adding to f a suitable diagonal f_1, we may suppose that

$$f \underset{2}{\sim} g^{(0)} + \phi^{(0)} + 4g^{(2)} + 4\phi^{(2)} + \dots, \qquad (222)$$

where for $r = 0, 2, 4, \dots$ we have

$$n(g^{(r)}) \text{ even}, \quad |A(g^{(r)})| \text{ odd}, \qquad (223)$$

and

$$n(\phi^{(r)}) = 2, \quad \phi^{(r)}(x_1, x_2) \equiv x_1^2 + 2x_1 x_2 \pmod 4. \qquad (224)$$

To justify (224), see (115) and (116). The binary forms 2-adically equivalent to $\pm (x_1^2 + cx_2^2)$, $c - 1$ or -3, which are excluded by the lemma just mentioned, are just those which on putting $x_1 + x_2$ for x_1 go into $\pm (x_1^2 + 2x_1 x_2 + 2x_2^2)$, modulo 4.

It will be convenient to regard (222) as a congruence modulo a suitable high power of 2. What (222) actually asserts is that f^P is congruent to the right-hand side modulo any power of 2 for some integral P with $|P|$ odd. Since, however, the right-hand side of (222) retains the shape (223) and (224) on putting hx_n for x_n, with any odd h, we may replace P by $P[1, \dots, 1, h]$, $= Q$, say, choose h so that $|Q| \equiv 1$, modulo the chosen power of 2, and then take $|Q|$ to be 1 by an argument used in the proof of Theorem 41.

With this we introduce the transformation $\mathbf{x} = M\mathbf{y}$ defined by $M = [2, \dots, 2, 1, \dots, 1]$, or by

$$x_i = 2y_i \quad \text{for} \quad i \leqslant 2h+1, \quad y_i \quad \text{for} \quad i > 2h+1, \qquad (225)$$

where $2h$ is the rank of $g^{(0)}$, and so $2h + 1 < n$ is the number of 2's in M, so that $|M| = 2^{2h+1} < 2^n$.

We shall first show that with this M we have $M^{-1}SM$ integral if S is integral. From (222) to (225) we see that the integral vector $\mathbf{x}$ satisfies the two congruences $2 | A\mathbf{x}$, $2 | f(\mathbf{x})$ simultaneously if and only if $\mathbf{x} = M\mathbf{y}$, $\mathbf{y}$ integral. Now since $f^S = f$ these two congruences are the same as $2 | S'AS\mathbf{x}$, $2 | f^S(\mathbf{x})$, or, since $|S'| = \pm 1$, as $2|ASx$, $2|f(Sx)$. They therefore hold if and only if $Sx = Mz$, z integral. Thus y is integral if and only if $z = M^{-1}SMy$ is so, whence the assertion.

With trivial modifications this argument shows that $w(M^{-1}SM)$ is odd if $w(S)$ is so. This means that the theorem is true for f if it is true for f^M; or for $\frac{1}{4}f^M$, which is clearly an integral form, and has discriminant numerically less than $d(f)$.

Now the effect of the transformation (225) is to leave all the terms on the right of (222) unaltered, except for $g^{(0)} + \phi^{(0)}$, which goes into $4g^{(0)} + \phi^{(0)}(2x_{2h+1}, x_{2h+2})$, $= 4g'$, say. It is clear that g' is an integral form with the properties (223), as is $g' + g^{(2)}$, $= g''$, say. Now it may be that the term $4\phi^{(2)}$ and subsequent terms in (222) are missing. If so, then $\frac{1}{4}f^M = g''$ has odd discriminant and the result follows.

In general, however, $\frac{1}{4}f^M$ satisfies

$$\frac{1}{4}f^M \underset{2}{\sim} g'' + \phi^{(2)} + 4g^{(2)} + \cdots,$$

and so is a form of the same shape as f. We can therefore repeat the process. Since the discriminant decreases at each step we ultimately get the result.

4. The norms of rational automorphs

We can now prove the key result:

THEOREM 81. *Every rational automorph S of a form f with discriminant d, with $w(S)$ prime to d, has the same norm as some product of rational reflexions of f, each reflexion having denominator prime to d.*

Proof. The theorem follows at once from Theorem 77 if d is odd (taking $m = d$); so we suppose d even. We may also suppose that f does not satisfy the hypothesis of Theorem 78, otherwise we could put $m = d$ in that theorem. This implies (see proof of Theorem 68) that $d_0 = d_0(f)$ defined in Chapter 7, §3 is odd.

We apply Theorem 77, with m the product of all the odd primes dividing d. Let $U(\mathbf{t}) = U(\mathbf{t}, f)$ be one of the factors in the product (220) so constructed. We shall have

$$f(\mathbf{t}) = 2^b aq, \quad w(U(\mathbf{t})) = 2^c q, \quad a|\mathbf{t}'A, \qquad (226)$$

for some integers a, b, c, q with

$$0 \leqslant c \leqslant b, \quad aq \text{ odd}, \quad a\,|\,d, \quad \text{g.c.d. } (q,d) = 1, \quad q > 0. \quad (227)$$

This is clear from the construction and Theorem 56.

Assuming as we clearly may that f is primitive, we can construct $\mathbf{z}$ (integral and primitive) with

$$2 \nmid f(\mathbf{z}), \quad f(\mathbf{z})f(\mathbf{t}) > 0, \quad \mathbf{z} \equiv \mathbf{t} \pmod{p^r} \quad (2 \neq p\,|\,d), \quad (228)$$

for a suitable sufficiently large r. It is clear that this gives us

$$w(U(\mathbf{z})) = aq', \quad q' > 0 \text{ odd}, \equiv 2^b q \pmod{p} \quad \text{for} \quad 2 \neq p\,|\,d. \quad (229)$$

Now carry out this construction for each of the reflexions in (220), introducing suffixes into the notation of (226)–(229) in the obvious way. We deduce from (226) and $2 \nmid w(S)$ that

$$w(S)\,q_1 \ldots q_u \quad \text{is a square;} \quad (230)$$

here we use the corollary to Theorem 55. From (229) and (230) we have

$$k^2 w(S)\,w(U(\mathbf{z}_1))\ldots w(U(\mathbf{z}_u)) \equiv 1 \pmod{d_0}, \quad (231)$$

for some integer k, if

$$b_1 + \ldots + b_u \equiv 0 \pmod{2}, \quad (232)$$

but

$$2k^2 w(S)\,w(U(\mathbf{z}_1))\ldots w(U(\mathbf{z}_u)) \equiv 1 \pmod{d_0} \quad (233)$$

if (232) is false. We use here $2 \nmid d_0$.

If (232) is false Theorem 80 shows that (221) is also false. Now part (i) of the lemma proved in Chapter 7, §2 shows that there exist reflexions U_1, U_2, each with denominator prime to d, such that $2w(U_1)\,w(U_2)$ is congruent to a square modulo d_0. In case (232) holds, whence also (231) holds, choose any U_1 with $w(U_1)$ prime to d, and take $U_2 = U_1$.

In either case, writing $U(\mathbf{z}_1) = U_3, \ldots$, it is clear from (231), (233) that for some k' we have

$$k'^2 w(S)\,w(U_1)\ldots w(U_{u+2}) \equiv 1 \pmod{d_0};$$

here, and for the rest of this proof, $U_1, U_2, \ldots$ are reflexions of f with denominators prime to d. The argument of Theorem 69 now shows that for some k'', v we have

$$k''^2 w(S)\,w(U_1)\ldots w(U_v) \equiv 1 \pmod{d}.$$

This shows that norm $(SU_1...U_v)$ is congruent to a square modulo d, and so by Theorem 60 equal to norm $U_{v+2}U_{v+1}$, for some U_{v+1}, U_{v+2}. Multiplying on the right by $U_{v+1}U_{v+2}$, and using $U^2 = I$ we find

$$\text{norm } (SU_1...U_{v+2}) = 1,$$

and by (171) this gives the desired result

$$\text{norm } S = \text{norm } (U_1...U_{v+2}).$$

The use that we can make of Theorem 81 has already been mentioned (see remark preceding Theorem 70). But we state and prove explicitly

THEOREM 82. *If f and f^R are both integral forms, $|R| = \pm 1$, and $w(R)$ is prime to $d = d(f) \neq 0$, then f, f^R are in the same spinor genus (hence, by Theorem 63, equivalent if they are indefinite and of rank at least 3) if and only if f has a set $U_1, U_2, ...$ of reflexions the product of whose weights is congruent to $w(R)$ modulo d.*

This congruence condition remains sufficient if weakened (i) *by putting $k^2w(R)$ (k any integer), or $\pm k^2w(R)$, if f is indefinite, for $w(R)$ and* (ii) *replacing the modulus d by $d_0 = d_0(f)$, satisfying $d_0^{\frac{1}{2}n(n-1)} | d$, but adding the requirement*

$$\text{g.c.d. } (w(U_1)w(U_2)..., d) = 1.$$

Proof. We start with the necessary and sufficient condition (199) of Theorem 69, that is, $k^2w(R)w(S) \equiv 1$ (or ± 1) (mod d_0), $w(S)$ prime to d. If this can be satisfied with $S = S_1$, and if norm $S_1 = \text{norm } S_2$, then it can also be satisfied, with possibly a different k, with $S = S_2$. So by Theorem 81 it can be satisfied with $S = U_1U_2...$, for some $U_1, U_2, ...$; and again altering the choice of k if necessary, we may replace $w(U_1U_2...)$ by $w(U_1)w(U_2)...$ (Theorem 55).

The condition is now $k^2w(R)w(U_1)w(U_2)... \equiv 1$ (or ± 1) (mod d_0), or with a different choice of k,

$$k^2w(R) \text{ (or } \pm k^2w(R)) \equiv w(U_1)w(U_2)....$$

It remains necessary with the sign $+$ in all cases, and also with the k^2 omitted; for if $U_1 = U(\mathbf{t})$, $\mathbf{t}$ primitive, we can replace U_1 by $U(\mathbf{z})$ with $\mathbf{z}$ primitive and $\mathbf{t} \equiv k\mathbf{z}$ (mod d). (This argument fails in the case of an empty set of reflexions $U_1, \ldots,$ which, however, though admissible, is always avoidable by using $U^2 = I$.)

The condition can be weakened still further but the results are complicated. It is easy to apply if the forms f, f^R in the same genus are given and a suitable R can be found.

5. Cayley's formula for the general rational automorph with determinant 1

We prove a simple and elegant—but not too useful—result due to Cayley:

THEOREM 83. *If* $A = A'$, *with* $|A| \neq 0$, *is the matrix of* $f = \frac{1}{2}\mathbf{x}'A\mathbf{x}$ *then the formula*
$$S = (A + Z)^{-1}(A - Z) \tag{234}$$

sets up a 1-1 *correspondence between rational skew matrices* Z *with* $|A + Z| \neq 0$ *and rational automorphs* S *of* f *with* $|I + S| \neq 0$; *and this last condition implies* $|S| = +1$.

Proof. Writing (234) as $(A + Z)(I + S) = 2A$, we see the need for the restrictions $|A + Z| \neq 0$, $|I + S| \neq 0$, and that rational Z, S satisfying them correspond one to one.

Now assume Z skew, that is, $Z' = -Z$, and (234). We deduce
$$S'(A + Z)S = S'(A - Z) = ((A + Z)S)' = (A - Z)' = A + Z.$$
Transposing, $S'(A - Z)S = A - Z$. Adding, $S'AS = A$, or $f^S = f$. Also $|A + Z| = |A + Z'| = |A - Z|$, and so $|S| = +1$.

Next assume $S'AS = A$, and (234). We have
$$Z(I + S) = A(I - S),$$
whence by transposition
$$(I + S')Z' = (I - S')A.$$
This gives
$$(I + S')(Z + Z')(I + S) = (I + S')A(I - S) + (I - S')A(I + S) = O,$$
and since $|I + S'| = |I + S| \neq 0, Z + Z' = O$ follows, that is, Z is skew, which completes the proof.

The exclusion of S with $|S| = -1$ is inconvenient for even n, as is the restriction $|I + S| \neq 0$ in all cases. The worst defect of the formula (234) is, however, that it is impossible to say what conditions on Z are needed if we wish to restrict the denominator of S in some way.

We can, however, make some interesting deductions from the fact that $|S| = -1$ implies $|I+S| = 0$. First, if n is odd we put $-S$ for S and see that $|S| = 1$ implies $|I-S| = 0$. Thus S has always an eigenvalue ± 1, $= |S|$; that is, there exists an eigenvector $\mathbf{x} \neq \mathbf{0}$, easily seen to be rational, with $S\mathbf{x} = |S|\,\mathbf{x}$. We could not, however, use this result as a substitute for the construction of Theorem 77, as it gives us no further information about the eigenvector. Similarly, if n is even and $|S| = -1$ then also $|-S| = -1$ and $|I+S| = |I-S| = 0$. That is, S has $1, -1$ as two of its eigenvalues.

For $n = 2$, this is enough to show that S must be a reflexion if $|S| = -1$; whence if $|S| = +1$ it follows that S is a product of two reflexions, one of which may be arbitrarily chosen. From this it is easy to verify that S is a linear combination, with rational scalar coefficients, of I and $\begin{pmatrix} \cdot & -1 \\ 1 & \cdot \end{pmatrix} A$. So we have

$$S = \tfrac{1}{2}tI + \tfrac{1}{2}u\begin{pmatrix} \cdot & -1 \\ 1 & \cdot \end{pmatrix} A, \tag{235}$$

where t, u are rational and the reason for the notation will appear shortly. Conversely, using the identity

$$R'\begin{pmatrix} \cdot & -1 \\ 1 & \cdot \end{pmatrix} R\begin{pmatrix} \cdot & -1 \\ 1 & \cdot \end{pmatrix} = -|R|\,I,$$

which is easily verified, we deduce from (235) that

$$S'AS = \tfrac{1}{4}(t^2 - du^2)A, \quad |S| = \pm\,\tfrac{1}{4}(t^2 - du^2).$$

This shows that (235) makes S an automorph of f, necessarily with $|S| = +1$, if and only if the rational numbers t, u satisfy the well-known PELLIAN EQUATION

$$t^2 - du^2 = 4. \tag{236}$$

6. Integral automorphs

We have made no use of integral automorphs, apart from occasional numerical examples. They are, however, of interest in themselves, and could be used for an alternative treatment (due to Meyer) of the problem of equivalence. For suppose that two forms in the same genus can be expressed in the shapes f^M, f^N, with integral M, N and the same f (with smaller discriminant than f^M, f^N). Then we may regard f^M as obtained from f by imposing on the variables x_i of f the congruence condition $(\text{adj } M)\mathbf{x} \equiv \mathbf{0} \pmod{|M|}$, this being the same as $\mathbf{x} = M\mathbf{y}$, $\mathbf{y}$ integral, because of $(\text{adj } M)M = |M|I$. It is therefore almost immediate that f^M, f^N are equivalent if f has an integral automorph S such that the congruence condition just mentioned goes into $(\text{adj } N)\mathbf{x} \equiv \mathbf{0} \pmod{|N|}$ on putting $S\mathbf{x}$ for $\mathbf{x}$. The converse is also true.

There is, however, no simple formula for the integral automorphs of a form with $n \geqslant 4$. For $n = 3$, they can be found from Hermite's formula (180); for it is not difficult to show that with g.c.d. $(x_1, ..., x_4) = 1$ this formula yields an integral S only if $x_4^2 - df(\mathbf{x})$ divides $4d^2$.

The integral automorphs of a definite form can always be found by calculation, and are finite in number; for $f^S = f$ implies $f(\mathbf{s}_j) = a_{jj}$, $\mathbf{s}_j$ being the jth column vector of S, and this gives finitely many possibilities for $\mathbf{s}_j$ with f definite. Integral reflexions clearly exist if and only if the form is equivalent to one with $a_{11} \mid a_{12}, ..., a_{1n}$. (Take the reflexion, by an equivalence transformation, to be $U(\{1, \mathbf{0}\})$, so that it takes $\mathbf{x}$ into

$$\{-x_1 - a_{11}^{-1}(a_{12}x_2 + ...), x_2, ...\}.)$$

Finally, let us consider the case $n = 2$, S integral; and in view of the last remark take $|S| = +1$. It is quite easy to see that S given by (235), for primitive f, is integral if and only if t, u are integers with $t \equiv a_{12}u \equiv du \pmod 2$; this condition is, however, implied by (236). In view of this, one could simplify (236) by putting $t, u = 2t_1 + dt_2, t_2$, giving

$$t_1^2 + dt_1t_2 + \tfrac{1}{4}(d^2 - d)t_2^2 = 1. \qquad (237)$$

The left member of (237) is a form with discriminant d, whereas that of (236) has discriminant $4d$. Unfortunately, however, this substitution complicates (235).

This leads to the problem of finding the solutions in integers of (236), or of (237). This problem has been very adequately treated elsewhere, and involves considerations outside the scope of this work.

NOTES

CHAPTER 1

§§ 1, 2. In the so-called classical notation, due to Gauss, binomial coefficients are introduced in the product terms. Thus, for example, $x_1 x_2$ is not regarded as an integral form, but as half of $2x_1 x_2$. In this notation the matrix of f is what is here called $\frac{1}{2}A$, and the determinant $\left|\frac{1}{2}A\right| = (-\frac{1}{4})^{[\frac{1}{2}n]} d$, takes the place of d. Unfortunately, this notation is still used (for $n \geqslant 3$) in much of the literature. In departing from it I follow Brandt (1945) and other modern German writers.

§ 4. I make no use of the relation of *proper equivalence* (defined as equivalence, with the further restriction $|T| = +1$); this relation lacks the very desirable property (17).

§ 5. I avoid confusion with old and convenient usage by not using the words *equivalence* and *class* in the way usual in abstract algebra, though, as the sophisticated reader will see, I regard an algebraic approach as helpful.

§ 7. Theorem 6 illustrates the use of the important property of an automorph S of f, that $f' = f^R$ implies $f' = f^{SR}$; by choice of S, SR may be made to satisfy certain conditions. Theorem 5, due to Witt, is quite a modern one, which for the logical development of the subject ought to have been known much earlier.

CHAPTER 2

§ 2. The minimum of an indefinite binary form can be found by using continued fractions; the inequalities (33) do not imply that it is equal to $|a_{11}|$. The proof of Theorem 8 shows that the form represents two integers of opposite signs whose product does not exceed $d/5$ in absolute value. Here the constant 5 can be replaced by 6 except in the special case (35). For further results (on indefinite binary forms with real coefficients) see Cassels (1957).

§ 3. To obtain further results we should have to study ideals in quadratic fields.

§ 5. The argument of Theorem 10 can be considerably sharpened for indefinite forms not representing zero. Suppose for example that $n = 3$, $|s| = 1$; then the binary form g with diagonal coefficients b_{22}, b_{33} is either indefinite or negative. Take the second case; the first is similar, though better handled by using the further result on binary forms mentioned in the above note on § 2.

The binary forms $f(x_1, x_2, 0)$ and $f(x_1, 0, x_3)$, say f_1, f_2, have discriminants $-b_{22}$, $-b_{33}$, and g has discriminant $16a_{11}d(f)$. Thus by Theorems 7 and 8 we have

$$|b_{22}| \geqslant 5a_{11}^2, \quad |b_{33}| \geqslant 5a_{11}^2, \quad 3\,|b_{22}b_{33}| \leqslant 16\,|a_{11}d(f)|.$$

These inequalities yield an improvement on $(48)_2$. If we can replace either of the 5's by 8 we get a better result. If not, Theorem 8 shows that f must be
$$a_{11}(x_1^2 + x_1 x_2 + x_1 x_3 - x_2^2 - x_3^2) + a_{23} x_2 x_3;$$
and this special case is easily dealt with.

For further results in this direction see Barnes and Oppenheim (1955) and Watson (1957 c, 1958), and papers there quoted. Most of these are on forms with *real* coefficients. A series of papers, mostly by Birch and Davenport, shows that if f has real coefficients and is indefinite, with rank at least 21, then $0 \leqslant f(\mathbf{x}) < \epsilon$ is soluble, with integral $\mathbf{x} \neq \mathbf{0}$, for every $\epsilon > 0$. I have stated this (see Theorem 15) so as to make it trivial for integral f; otherwise it is completely outside the scope of this work, though of great interest. See Ridout (1958).

§ 8. See Mordell (1944) and Oppenheim (1953). The result assumed in the first example was proved (with the better constant $\frac{1}{10}$ except for $f/\min f \sim \pm (x_1^2 + x_1 x_2 + x_2^2) \mp 2x_3^2$) by Oppenheim.

§ 11. The suggested method for calculating the bound is due to Korkine and Zolotareff and practicable for $n \leqslant 5$. For a deeper method, due to Voronoï, see Barnes (1957) (for the case $n = 6$; the method is general but the labour increases rapidly with n). Blichfeldt calculated the bound for $n \leqslant 8$, using the Hermite reduction. It is not known for $n \geqslant 9$, though Blichfeldt has obtained asymptotic results for large n. The present state of knowledge is summarized by
$$|d|/\min{}^n f \geqslant 3, 2, 4, 2, 3, 1, 1 \quad \text{for} \quad n = 2, 3, 4, 5, 6, 7, 8, \ f \text{ definite;}$$
each bound is best possible.

CHAPTER 3

§ 1. By using the theory of p-adic numbers, the infinite system of congruences (73) can be replaced by a single equation. There would be no substantial advantage in using this theory, but we borrow from it the convenient term p-*adic*.

§ 3. I base this account of the Hilbert symbol mainly on the simplified treatment of the symbol given by Pall (1945), except for the introduction of ternary p-adic zero forms, which seems to me to be useful.

§ 4. For a fuller treatment of quadratic residuacity (without the Hilbert symbol) see, for example, Hardy and Wright. Burgess (1957) gives the best known estimate for the least positive q with $(q \mid p) = -1$ when p is a large prime.

§ 5. The result here proved for $n \geqslant 5$ is a classical one, due to Meyer. It will be noted that we have now a method for determining, by a strictly finite calculation, whether or not a given form represents zero, but no practical method of finding a representation. Cassels (1955, 1956; see also Davenport) has found inequalities satisfied by the simplest representation. The problem of solving simultaneous equations $f_1(\mathbf{x}) = f_2(\mathbf{x}) = 0$ has been recently tackled by Professor Mordell.

§ 6. Again see Pall (1945), and earlier work by Hasse and others there quoted. Pall's invariant $c_p(f)$ is, in my notation,

$$(-1, -1)_p \epsilon_p(f) \quad \text{for} \quad n \equiv 2, 3 \, (\text{mod} \, 4),$$
$$(-d, -1)_p \epsilon_p(f) \quad \text{for} \quad n \equiv 0, 1 \, (\text{mod} \, 4).$$

§ 7. It is of interest to construct the form f of Theorems 26 and 27 so that $|d(f)|$ is as small as possible (see Brandt (1945)).

CHAPTER 4

§ 1. As the definition indicates, we do not, in this chapter, allow p to take the value ∞; though it would be possible to define ∞-adic equivalence to mean equality of rank and of signature. The sophisticated reader will observe that whereas in the last chapter we worked in the p-adic field, in this we work in its integer ring; this distinction disappears for $p = \infty$.

§ 2. In the Gaussian notation, the analogue of (99) or (100) is valid only for odd p, and may be written (for $p \nmid 2d$) as $f \underset{p}{\sim} x_1^2 + \ldots + x_{n-1}^2 + \Delta x_n^2$, $p \nmid \Delta$. The results on solutions of congruences become considerably harder to state and to prove.

§ 3. Theorem 32 shows that the converse of (18) holds when $\mathscr{R}$ is p-adic equivalence, $p \neq 2$; for the p-adic invariants of $f_1 + f_2$ can obviously be found from those of f_1, f_2 by adding the n_r and multiplying the $(d_r | p)$ ($= \pm 1$), and these processes are reversible. It is clear from the example (29) that this is not so for $p = 2$.

§ 5. We depart from the 'classical' theory by splitting off as *many* binary forms as possible. For both $x_1 x_2$ and $x_1^2 + x_1 x_2 + x_2^2$ have simpler congruence properties than x_1^2.

§ 7. These results can be used (Watson, 1955, 1957 c) to obtain a good estimate for the least positive a for which all the necessary congruence conditions for a to be (properly) represented by f hold.

§ 8. The argument shows that if we can take out a given prime p at all we can do so without putting other primes into the discriminant. Hence one can prove the results of Brandt mentioned in the note of Chapter 3, § 7.

CHAPTER 5

§ 1. For the earliest use, as far as I know, of this relation, see Watson (1958).

§ 2. It simplifies the argument not to have to consider the signature, as we should have to do if we dealt with semi-equivalence without the preliminary of congruential equivalence.

§ 3. The results are in substance classical, and the definition is substantially that of Minkowski. But I believe I can claim priority for noticing that the necessary and sufficient conditions for the existence of a genus with specified properties are unaffected if the specified properties are varied merely by altering the signature, provided that its residue modulo 8 is unaltered (and subject to $|s| \leqslant n$). The treatment is based in part on that of Pall (1945).

§ 4. I believe these results are original; the general problem of decomposition has not received the attention it deserves. The constants 4, 8, 7, 11 are all best possible. It is somewhat tedious to prove this, but a simple example will show that a substantial improvement is not possible. Suppose $n(\phi) = s(\phi) = 8$, $d(\phi) = 1$ (see note on Chapter 3, § 11 for the existence of such a ϕ). Suppose that $f = \phi + \phi + \ldots + \phi$ has a decomposition $f \simeq f_1 + f_2$, with $n(f_1) = n_1$. Then (9) shows that n_1 is even and that $d_1 = d(f_1) \mid d(f) = 1$, $d_1 = \pm 1$. Now (52) gives $d_1 = 1$, and as f_1 must be positive (23) gives $4 \mid n_1$. $d_1 = 1$ gives, using (92) and (95), $\epsilon_\infty(f_1) = 1$; whence $8 \mid n_1$ by (91).

§ 5. The Eisenstein–Smith definition of the genus is in substance that $f \simeq f'$ if there exists a rational R with denominator prime to $2d$, and determinant 1, such that $f^R = f'$. Theorem 50 (in which it is easy to see that we could make the ambiguous sign $+$) shows (i) that this definition is equivalent to that of Minkowski and (ii) that it can be simplified by writing d for $2d$.

§ 6. If there is only one class in the genus of f, then the f' of Theorem 51 is equivalent to f; so f itself represents (properly or otherwise) all integers satisfying the generic conditions. This is the case for many genera with small n, s, d; for example if $n = s = 3$ and $|d| \leqslant 10$. In this way some well-known elementary results can be obtained. For example, the form $x_1^2 + x_2^2 + x_3^2$ represents $a > 0$ properly if and only if $4 \nmid a$ and $a \not\equiv -1$ (mod 8) (Legendre's three-square theorem). Similarly, $x_1^2 + \ldots + x_4^2$ represents properly $a > 0$ if and only if $8 \nmid a$. (Here, as usually for $n \geqslant 4$, there is an alternative proof by induction on n.)

§ 7. It is surprising that Theorem 53 should not have been proved till quite recently. See Watson (1955); but Siegel (1951) gives an earlier (and more difficult) proof of a more general result.

CHAPTER 6

§ 1. See MacDuffee (1946) for a treatment of this problem under more general conditions.

§ 2. Theorem 54 is proved in Watson ($1957d$); it seems to have previously escaped notice.

§ 3. I believe Theorem 55 is new. Without it, the norm of a rational matrix is not useful. I have been unable to find a simpler proof.

§ 4. The reflexions (and other rational automorphs) of a quadratic form have been studied by many writers; see, for example, Jones and Marsh (1954), and references there given.

§ 5. For a fuller treatment of Hermite's formula see Jones and Watson (1956); further results are there given, which we shall not here need. In its original shape, without modern matrix notation, the formula is very unwieldy.

§ 7. These results can be taken further, but this is best done after introducing the spinor genus, which we do in the next chapter.

CHAPTER 7

§ 1. Much of this chapter is essentially a specialization of the deep work of Eichler (1952). The forms f, f' might be said to be *properly* spinor-related if the conditions of this definition can be satisfied with $|R| = +1$; if so, they can be shown to be *spinor-verwandt* according to Eichler's definition.

Theorem 63 (due to Eichler) takes the theory of equivalence as far as we can go. A spinor genus of definite forms may contain arbitrarily many classes. It may, however, be worth while to remark that one can always determine by a finite calculation whether two given positive forms are equivalent or not. For $f'^T = f$ gives $f'(\mathbf{t}_j) = a_{jj}$, where $\mathbf{t}_j$ is the jth column vector of T, and this gives only finitely many possibilities for $\mathbf{t}_j$.

§ 4. It is clear that the assertion of Theorem 71 is false, except in the case in which the genus contains only one spinor genus, if the ϕ_i are all unary. But by arguments like those of § 6 it can be shown to be true in general if some ϕ_i is binary. For example, it is true if $n(\phi_1) = n(\phi_2) = 2$ and $d(\phi_1)\,d(\phi_2)$ is not a square.

§ 5. If $n \geqslant 5$, the argument of Theorem 72 can be modified so as to take f^R rationally into a form in the spinor genus of f while leaving $f^R(x_1, \ldots, x_{n-3}, 0, 0, 0)$ unaltered. That is, if some form in the genus of f represents properly a given $(n-3)$-ary form f_1, then some form in the spinor genus of $f(f$ itself if indefinite) represents f_1 properly (i.e. reduces to f_1 on making a suitable equivalence transformation and then putting three of the variables equal to 0).

§ 6. This section deals with a problem first investigated by Meyer in the 1890's. The number of possible values for the number a_0 may be shown to be one less than the number of spinor genera in the genus. It may be however that no a satisfying (206), with given a_0, is exceptional even though a_0 satisfies all the conditions of Theorems 73 and 74. Meyer's results (for the indefinite case only) are quoted and extended a little in Jones and Watson (1956).

§ 7. Theorem 76 is due mainly to Kloosterman; but see Ross and Pall (1946) for references, and for removal of Kloosterman's restriction to diagonal forms. The results are stated for representations not restricted to be proper, but it is quite straightforward to get rid of the improper ones. The form $2(x_1^2 + \ldots + x_{n-1}^2) + (2b+1)\,x_n^2$ is clearly (Theorems 66 and 67) in a genus with only one spinor genus, but with at least two classes. This proves that the condition $n > |s|$ in Theorem 63 is necessary.

Positive ternary forms have been investigated by Linnik, but I do not know whether Theorem 76 would be true for $n = 3$ if exceptional a were excluded. Without this modification it is of course false for $n = 3$, as the example with $d = -27$ in § 5 shows.

Dickson (1939) investigated *regular* positive ternary forms, that is, positive ternary forms representing all the integers represented by their respective genera. I carried his work further in an unpublished thesis. The number of classes of such forms is large but finite; and their regularity can be proved by arguments of the kinds given in this tract.

CHAPTER 8

§§ 1, 2. The problem here studied does not seem to have previously received the attention it deserves, though there is a considerable literature on the simpler problem of factorization into reflexions with unrestricted denominators.

§ 3. Theorem 79 remains true if the modulus signs in the definition of the integer q are omitted; but the proof needs difficult algebra, not needed for any of the main results of this work. With this modification, q (possibly negative) is essentially the norm of S according to Eichler's definition.

§ 4. The application of Theorem 82 may present a little difficulty if we are given two forms f, f' in the same genus and so have to find an R satisfying the conditions in order to find whether f, f' are spinor-related. It is clear that there exists a simplest such R with bounded denominator and bounded elements; explicit bounds would be of interest. It might be better to find first an R with unrestricted denominator, and then multiply it by reflexions so as to obtain a new R with $w(R)$ prime to d.

§ 5. There are other formulae for the general rational automorph; for example, using the same sort of algebra as mentioned in the above note on § 3, one can express it (see, for example, Eichler (1952)) in terms of generalized quaternions.

§ 6. For an interesting property of the integral automorphs of a form not representing zero, and an application, see Watson (1957 a).

REFERENCES

BARNES, E. S. (1957). The complete enumeration of extreme senary forms. *Phil. Trans.* A, **249**, 461–506.

BARNES, E. S. and OPPENHEIM, A. (1955). The non-negative values of a ternary quadratic form. *J. Lond. Math. Soc.* **30**, 429–39.

BIRCH, B. J. and DAVENPORT, H. (1958). Quadratic equations in several variables. *Proc. Camb. Phil. Soc.* **54**, 135–8.

BRANDT, H. (1945). Über quadratische Kern- und Stammformen. Festschrift zum 60. Geburtstag von Prof. Dr Andreas Speiser. Zürich, Orell Füssli Verlag.

BRANDT, H. (1952). Über Stammfaktoren bei ternären quadratischen Formen. *Berichte über die Verhandlungen der sächsischen Akademie der Wissenschaften zu Leipzig, Mathematisch-naturwissenschaftliche Klasse*, **100**, 1–24.

BURGESS, D. A. (1957). The distribution of quadratic residues and non-residues. *Mathematika*, **4**, 106–12.

CASSELS, J. W. S. (1955, 1956). Bounds for the least solutions of homogeneous quadratic equations. *Proc. Camb. Phil. Soc.* **51**, 262–4 and **52**, 604.

CASSELS, J. W. S. (1957). *An Introduction to Diophantine Approximation* (Cambridge Tract no. 45).

DICKSON, L. E. (1918). *History of the Theory of Numbers*, vol. III, *Quadratic and Higher Forms.* New York, 1952.

DICKSON, L. E. (1939). *Modern Elementary Theory of Numbers.* Chicago.

EICHLER, M. (1952). *Quadratische Formen und orthogonale Gruppen.* Berlin.

HARDY, G. H. and Wright, E. M. (1938). *An Introduction to the Theory of Numbers.* Oxford.

JONES, B. W. and MARSH, D. (1954). Automorphs of quadratic forms. *Duke Math. J.* **21**, 179–93.

JONES, B. W. and WATSON, G. L. (1956). On indefinite ternary quadratic forms. *Canad. J. Math.* **8**, 592–608.

KNESER, M. (1956). Klassenzahlen indefiniter quadratischer Formen in drei oder mehr Veränderlichen. *Arch. Math.* **7**, 323–32.

MACDUFFEE, C. C. (1946). *The Theory of Matrices.* Ergeb. der Math., New York.

MORDELL, L. J. (1944). Observation on the minimum of a positive quadratic form in eight variables. *J. Lond. Math. Soc.* **19**, 3–6.

OPPENHEIM, A. (1953). Values of quadratic forms (I, II). *Quart. J. Math.* (Oxford), **4**, 54–9, 60–6.

PALL, G. (1935). On the order invariants of integral quadratic forms. *Quart. J. Math.* (Oxford), **6**, 30–51.

PALL, G. (1945). The arithmetic invariants of quadratic forms. *Bull. Amer. Math. Soc.* **51**, 185–97.

RANKIN, R. A. (1956). On the minimal points of positive definite quadratic forms. *Mathematika*, **3**, 15–24.

RIDOUT, D. (1958). Indefinite quadratic forms. *Mathematika*, **5**, 122–4.

ROSS, A. E. and PALL, G. (1946). An extension of a problem of Kloosterman. *Amer. J. Math.* **68**, 59–65.

SELBERG, A. (1949). An elementary proof of Dirichlet's theorem about primes in an arithmetic progression. *Ann. Math.* (2), **50**, 297–304.

SIEGEL, C. L. (1951). Indefinite quadratische Formen und Funktionentheorie. I. *Math. Ann.* **124**, 17–54.

WATSON, G. L. (1954). The representation of integers by positive ternary quadratic forms. *Mathematika*, **1**, 104–10.

WATSON, G. L. (1955). Representation of integers by indefinite quadratic forms. *Mathematika*, **2**, 32–8.

WATSON, G. L. (1957a). Bounded representation of integers by quadratic forms. *Mathematika*, **4**, 17–24.

WATSON, G. L. (1957b). Least solutions of homogeneous quadratic equations. *Proc. Camb. Phil. Soc.* **53**, 541–3.

WATSON, G. L. (1957c). The minimum of an indefinite quadratic form with integral coefficients. *J. Lond. Math. Soc.* **32**, 503–7.

WATSON, G. L. (1957d). The equivalence of quadratic forms. *Canad. J. Math.* **9**, 526–48.

WATSON, G. L. (1958). One-sided inequalities for integral quadratic forms. *Quart. J. Math.* (Oxford), **29**, 99–108.

INDEX